COMPUTATION AND THEORY IN ORDINARY DIFFERENTIAL EQUATIONS

A SERIES OF BOOKS IN MATHEMATICS

Editors: R. A. ROSENBAUM
G. PHILIP JOHNSON

COMPUTATION AND THEORY IN ORDINARY DIFFERENTIAL EQUATIONS

JAMES W. DANIEL
RAMON E. MOORE

University of Wisconsin

W. H. FREEMAN AND COMPANY
SAN FRANCISCO

Printed in the United States of America

Library of Congress Catalog Card Number: 71-117611
International Standard Book Number: 0-7167-0440-4
AMS 1970 Subject Classifications 34A50, 65L05, 65L10.

1 2 3 4 5 6 7 8 9

to our wives, Linda and Judy

PREFACE

Some of this material was first presented by one of the authors (R.E.M.) as a one-semester graduate level topics course in numerical analysis during the spring of 1966 in the Computer Sciences Department at the Madison campus of the University of Wisconsin; the present text represents an expanded version of the lecture notes for that course. Our intent is to present briefly the theoretical background in differential equations and the modern computational methods of solving such equations in such a way as to emphasize the interaction between the two topics. Thus, this book is not a traditional text in either the theory of differential equations or numerical analysis alone; no theorems are stated and proved in detail. Our primary goal is to indicate some ways in which theoretical concepts and computation can be combined to attack a problem efficiently.

With this goal in mind, we have divided the material into three parts. The first two parts comprise a survey of abstract theory and computational methods that use theoretical insights. In the third part we examine transformations that can be used as analytical and computational tools to solve many kinds of problems; it is one of our main contentions that pre-analysis and transformations which utilize partial knowledge about the solutions are of great value. We are primarily interested in initial value problems, especially in Part 3, although we do examine boundary value problems as well; eigenvalue problems are not included at all.

Although much of this text could be understood by a person without knowledge of either differential equations or numerical analysis, it is intended for an audience aware of the nature of both of these subjects and interested in studying their interactions in order to achieve more accurate and efficient computation. We hope that this material will help raise interest in the general application of analytical techniques to computations; we would especially like to see the growth of machine-automated analytical methods, which have only recently become feasible.

Thanks are due—and are herewith given to—our colleague, Ben Noble, whose many suggestions are invaluable.

Madison, Wisconsin
November, 1969

James W. Daniel
Ramon E. Moore

CONTENTS

COMPUTATION AND THEORY IN ORDINARY DIFFERENTIAL EQUATIONS

NOTATIONAL CONVENTIONS

A bibliography of books and articles cited in the text begins on page 165: throughout the text these are referred to by the numbers given them in the bibliography. References to parts of this text itself take the form "Section 2.3," which refers to the third titled section in Chapter 2, or the form "Eq. 2–3." which refers to the third numbered equation in Chapter 2.

Vectors will be denoted both as $(y_1, \ldots, y_n)$ and $\begin{pmatrix} y_1 \\ \vdots \\ y_n \end{pmatrix}$. If $\mathbf{y}$ is a vector and $\mathbf{f}$ a vector-valued function of $\mathbf{y}$, $\mathbf{f}(\mathbf{y})$ denotes $(f_1(y_1, \ldots, y_n), \ldots, f_n(y_1, \ldots, y_n))$. The matrix $\partial \mathbf{f}/\partial \mathbf{y}$ has as its i, j element the scalar $\partial f_i/\partial y_j$.

MATHEMATICAL SYMBOL	MEANING
E^n	n-dimensional real Euclidean space
$\\|\cdot\\|$	An arbitrary vector norm and the generated matrix norm
$\\|\cdot\\|_\infty$	The maximum norm For vectors: $\\|\mathbf{x}\\|_\infty = \max_{1 \leq i \leq n} \|x_i\|$ For matrices: $\\|\mathbf{A}\\|_\infty = \max_{1 \leq i \leq n} \sum_{j=1}^{n} \|A_{ij}\|$
$\in$	Is a member of
$\subset$	Is a subset of
$'$ (prime)	$\dfrac{d}{dt}$
$u(h) = O(h)$	There is a constant K such that $\|u(h)\| \leq K\|h\|$ as $h \to 0$
$u(h) = o(h)$	$\lim_{h \to 0} \left(\dfrac{u(h)}{h}\right) = 0$

PART 1

THEORETICAL BACKGROUND

Many mathematical problems that arise in practice do not fit into neat classes that can be treated by a well-developed theory. A reasonable man, however, should make use of any applicable theory whenever possible in order to acquire some understanding of the behavior of the solution to his problem. In these first three chapters we survey briefly some of the theoretical results that we feel can most often contribute to reducing the overall effort of computing a solution. We by no means claim to review everything that might be useful.

Chapter 1 discusses geometric concepts related to differential equations, and therefore dwells briefly on such topics as curves, surfaces, trajectories, integral surfaces, and vector fields. Chapter 2 examines several aspects of initial value problems, including the existence, uniqueness, and representation of solutions; dependence of the solutions on the initial data and on the differential equation (perturbation theory); and the asymptotic behavior of solutions. The chapter concludes with a few remarks on periodic solutions.

Existence, uniqueness, and representation of solutions to boundary value problems are treated in Chapter 3, along with reformulations of the problem as an integral equation or variational problem; we also discuss a relevant monotonicity property.

1

BASIC CONCEPTS; GEOMETRY

1.1 INTRODUCTION

We are interested in ordinary differential equations, that is, in equations involving the derivative of an unknown function with respect to a single variable t, often understood to represent time. In particular, we often will consider equations of the form

$$\mathbf{y}' = \mathbf{f}(\mathbf{y}).$$

If $y = y(t)$ is a real-valued function of the real variable t, the differentia equation tells y whether it should increase ($y' = f(y) \geq 0$) or decrease ($y' = f(y) \leq 0$), and how rapidly it should do either (rapidly if $|y'| = |f(y)|$ is large) at any value of y.

More generally, if $\mathbf{y}$ and $\mathbf{f}$ are vectors in n-dimensional Euclidean space E^n, so that $\mathbf{y} = (y_1, \ldots, y_n)$ and $\mathbf{f}(\mathbf{y}) = (f_1(y_1, \ldots, y_n), \ldots, f_n(y_1, \ldots, y_n))$, the differential equation still describes how at any point $\mathbf{y}$ the vector $\mathbf{y}$ must modify itself in order to qualify as a solution. Just as in one dimension a positive or negative value for $f(y)$ means "increase" or "decrease," the vector $\mathbf{f}(\mathbf{y})$ indicates the direction in which $\mathbf{y}$ must move, and the magnitude of $\mathbf{f}(\mathbf{y})$ indicates speed. Such an intuitive geometrical view of differential equations is often useful; we pursue these notions further.

1.2 VECTOR FIELDS; DIFFERENTIABLE CURVES AND SURFACES

Let $\mathbf{Y}$ be a set of vectors $\mathbf{y} = (y_1, \ldots, y_n)$ in E^n. A *vector field* on $\mathbf{Y}$ is a mapping $\mathbf{f} = (f_1, \ldots, f_n)$ assigning a vector $\mathbf{f}(\mathbf{y}) \in E^n$ to each $\mathbf{y} \in \mathbf{Y}$. The vector field is said to be continuous if the mapping **f** is continuous; this is equivalent to continuity for each of the real-valued functions $f_i(\mathbf{y})$, $i = 1, 2, \ldots, n$. Perhaps the simplest way to describe a vector field graphically is

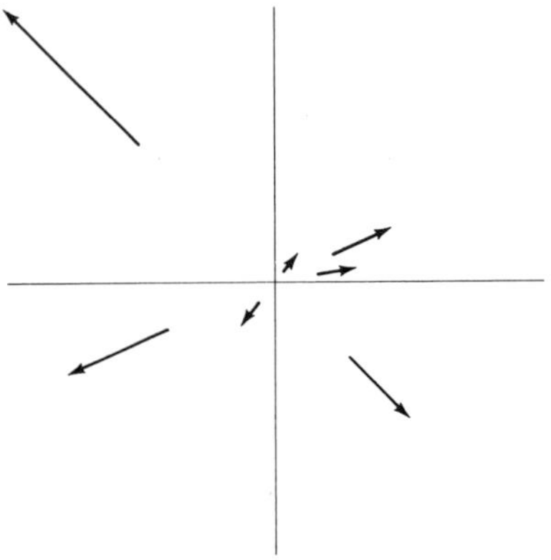

FIGURE 1–1. $\mathbf{f}(y_1, y_2) = (y_1, y_2)$

to draw the vector $\mathbf{f}(\mathbf{y})$ as emanating from the point **y**. Thus, Figures 1–1 through 1–4 depict vector fields on E^2, although not all possible vectors are drawn.

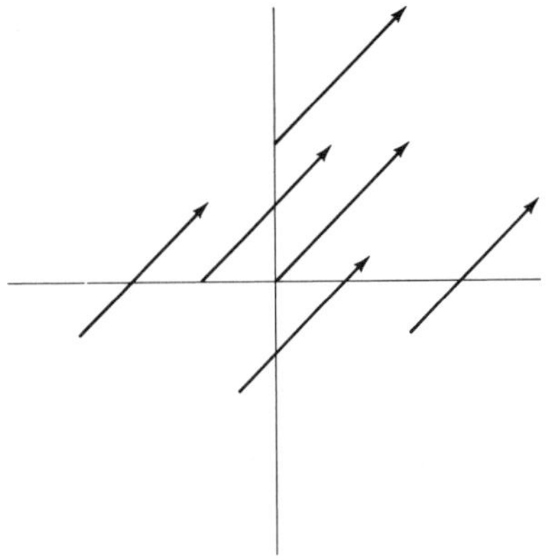

FIGURE 1–2. $\mathbf{f}(y_1, y_2) = (1,1)$

The first three vector fields are continuous; the fourth is not, since $f_1(y_1, y_2)$ is discontinuous along the line $y_1 = 1$. The third vector field looks somewhat discontinuous along the unit circle because of the directions of the field near there; the limit from inside or outside the circle is the same, however, saving the day. The limit is, in fact, zero (i.e., (0, 0)), and the field is said to *vanish* at such a point. The field in Figure 1–1 vanishes only at $\mathbf{y} = (0, 0)$; that in Figure 1–2 never vanishes; that in Figure 1–3 vanishes at $y = (0, 0)$ as well as along $y_1^2 + y_2^2 = 1$; and that in Figure 1–4 vanishes along $y_1 = 0$.

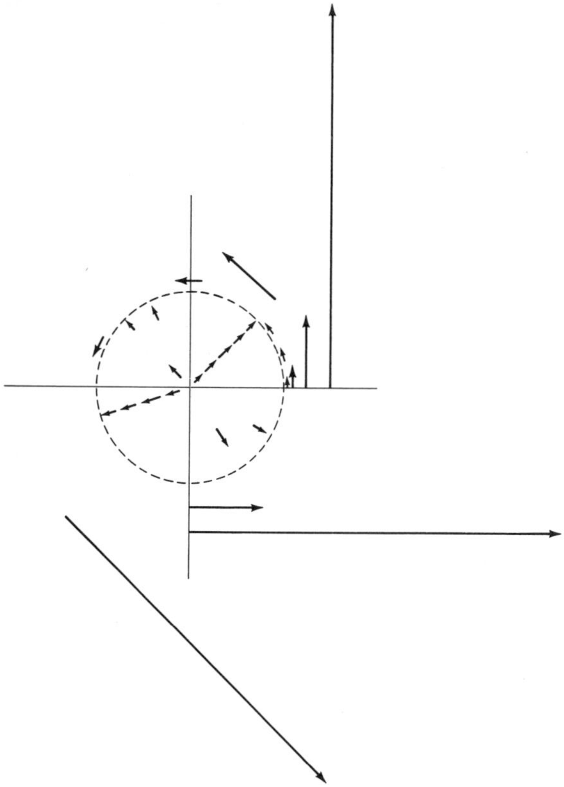

FIGURE 1–3. $\mathbf{f}(y_1, y_2) = (y_2(1 - y_1^2 - y_2^2), -y_1(1 - y_1^2 - y_2^2))$ if $y_1^2 + y_2^2 \geq 1$,
$= (y_1(1 - y_1^2 - y_2^2), y_2(1 - y_1^2 - y_2^2))$ if $y_1^2 + y_2^2 \leq 1$.

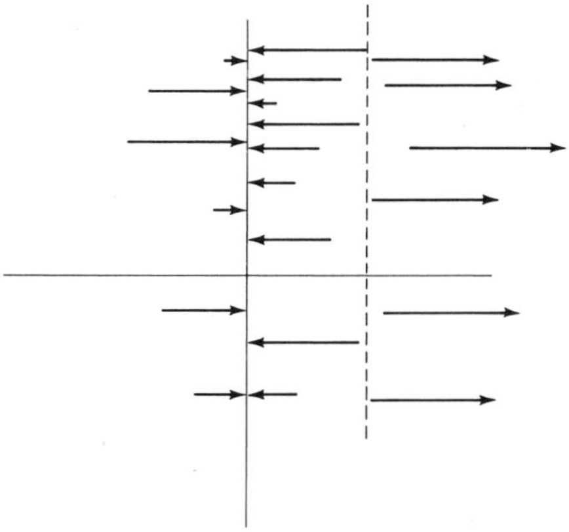

FIGURE 1–4. $\mathbf{f}(y_1, y_2) = (y_1, 0)$ if $y_1 > 1$
$= (-y_1, 0)$ if $y_1 \leq 1$

EXERCISE

Verify the statements concerning Figures 1–1 to 1–4.

A solution to a differential equation is a vector-valued function $\mathbf{y} = \mathbf{y}(t)$ in E^n; as such, the vector $\mathbf{y}$ defines a curve in E^n as t varies. More precisely, a *curve* in E^n is the set of values $\mathbf{y}(t)$ of a continuous vector function $\mathbf{y}$ as t varies over an interval I of the real line; the interval I may be open, half-open, closed, and bounded or unbounded on either end. The curve is *differentiable* if the function $\mathbf{y}(t) = (y_1(t), \ldots, y_n(t))$ is differentiable on I; if a curve defined by $\mathbf{y}$ is differentiable, then at any time t_0 the vector $\mathbf{y}' = (y_1'(t_0), \ldots, y_n'(t_0))$ is *tangent* to the curve, that is, instantaneously the curve moves in the direction $\mathbf{y}'$ from the point $\mathbf{y}(t_0)$. Figures 1–5 and 1–6 show curves in E^2.

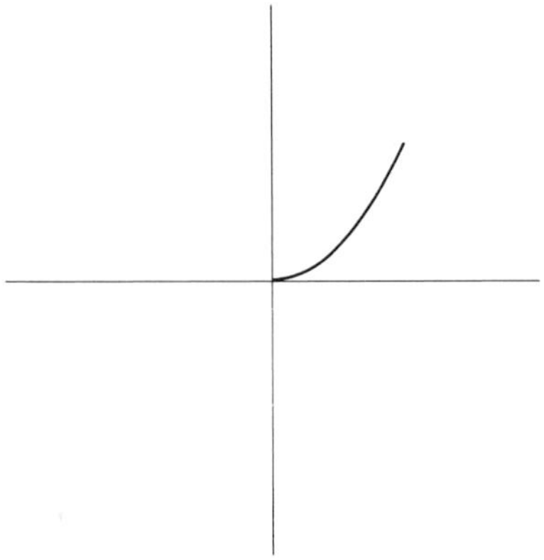

FIGURE 1–5. $\mathbf{y}(t) = (t, t^2), I = [0, 1]$

EXERCISE

Plot the curve $\mathbf{y}(t) = (te^t, t + \ln t)$.

The final geometric concept we wish to recall is that of a surface in E^n. A set $\mathbf{S}$ in E^n is a *surface* if there exists a differentiable real-valued function $s(\mathbf{y})$ defined on E^n such that $\mathbf{S} = \{\mathbf{y};\ s(\mathbf{y}) = 0\}$; intuitively the set $\mathbf{S}$, being subject to one constraint, is "$(n - 1)$-dimensional."

Additional references: 10, 24, 38, 43, 61, 74, 107.

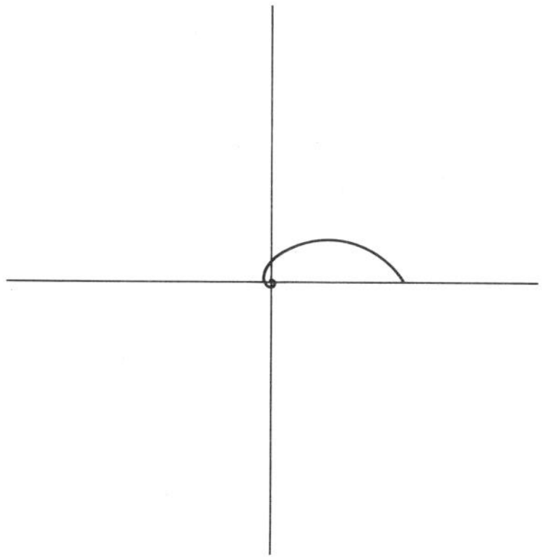

FIGURE 1–6. $\mathbf{y}(t) = (e^{-t}\cos t, e^{-t}\sin t)$, $I = [0, \infty]$

1.3 DIFFERENTIAL EQUATIONS; TRAJECTORIES; INTEGRAL SURFACES

Let $\mathbf{f}(\mathbf{y})$ define a vector field on E^n (or perhaps on a subset Y of E^n). The equation

$$\mathbf{y}' = \mathbf{f}(\mathbf{y})$$

for an unknown function $\mathbf{y}(t)$ is called an *autonomous* system of ordinary differential equations; "autonomous" refers to the fact that $\mathbf{y}(t)$ is not explicitly governed by time in the equation. If one has the time-dependent system $\mathbf{x}' = \mathbf{g}(t,\mathbf{x})$ for $\mathbf{x} \in E^n$, it can be converted into an autonomous system by the ploy: let $\mathbf{y} \in E^{n+1}$, $y_1 = x_1, \ldots, y_n = x_n$, $y_{n+1} = t$, let $\mathbf{f}(\mathbf{y}) = (f_1(y_1, \ldots, y_{n+1}), \ldots, f_{n+1}(y_1, \ldots, y_{n+1})) = (g_1(y_{n+1}, y_1, \ldots, y_n), \ldots, g_n(y_{n+1}, y_1, \ldots, y_n), 1)$, and then we have $\mathbf{y}' = \mathbf{f}(\mathbf{y})$. Computationally, one would work with $\mathbf{x}$ and $\mathbf{g}$ rather than with $\mathbf{y}$ and $\mathbf{f}$, but for purposes of geometric insight the autonomous form is preferable. For an example of an autonomous system, let

$$\mathbf{f}(\mathbf{y}) = \mathbf{A}\mathbf{y},$$

where $\mathbf{A}$ is an n by n matrix of constants; then the equation

$$\mathbf{y}' = \mathbf{A}\mathbf{y}$$

has the general solution

$$\mathbf{y}(t) = e^{t\mathbf{A}}\mathbf{y}(0),$$

where $\mathbf{y}(0)$ is arbitrary and

$$e^{t\mathbf{A}} = I + t\mathbf{A} + \frac{t^2}{2}\mathbf{A}^2 + \ldots$$

converges absolutely for each t and any $\mathbf{A}$. Applying this to the equation

$$x'' + k^2 x = 0$$

for the simple harmonic oscillator, setting $\mathbf{y} = (x, x')$, we get $\mathbf{y}' = \mathbf{A}\mathbf{y}$, where

$$\mathbf{A} = \begin{pmatrix} 0 & 1 \\ -k^2 & 0 \end{pmatrix}$$

and

$$e^{t\mathbf{A}} = \begin{pmatrix} \cos kt & \dfrac{\sin kt}{k} \\ -k \sin kt & \cos kt \end{pmatrix}.$$

In particular, this yields

$$x(t) = x(0)\cos kt + \frac{x'(0)}{k} \sin kt.$$

EXERCISE

Verify the two-by-two expression for $e^{t\mathbf{A}}$.

Since a solution to $\mathbf{y}' = \mathbf{f}(\mathbf{y})$ must certainly be piecewise differentiable, $\mathbf{y}$ defines a piecewise differentiable curve C. The tangent to C at time t is $\mathbf{y}' = \mathbf{f}(\mathbf{y})$, so the vector field $\mathbf{f}(\mathbf{y})$ gives us the tangent vector to *any* solution curve passing through any point $\mathbf{y}$ in E^n. Hence, $\mathbf{y}$ can be thought of as a *trajectory* whose instantaneous behavior at any point $\mathbf{y}$ is determined by $\mathbf{f}(\mathbf{y})$. For autonomous systems in E^2 for which two-dimensional paper is sufficient for graphing purposes, even the most clumsy artist can sketch tentative trajectories, given the vector field $\mathbf{f}(\mathbf{y})$; see Figure 1–7 for an example. In higher dimensions, the lack of high-dimensional graph paper somewhat restricts us, but much geometric insight is still available through this use of trajectories.

EXERCISE

Sketch trajectories for $\mathbf{f}(y_1, y_2) = (y_2, -y_1)$.

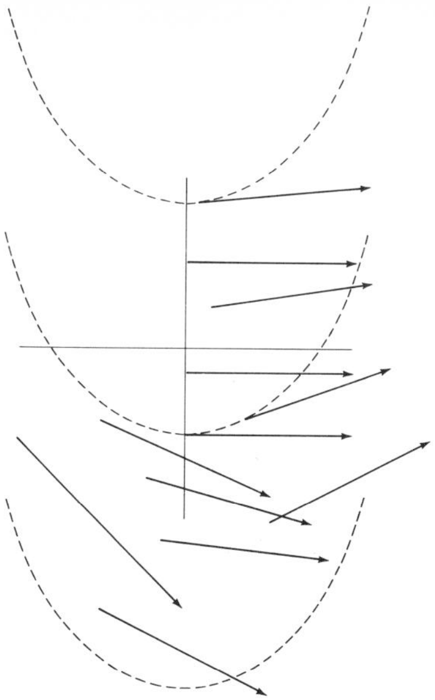

FIGURE 1–7. $\mathbf{f}(y_1, y_2) = (2, y_1)$. Vector field: arrows. Trajectories: dashes.

The reader should not be misled, however, into thinking that trajectories are always uncomplicated-looking curves in E^n. Consider, for example, the equation in E^3:

$$y_1' = -y_2 - \frac{\alpha y_1 y_3}{\sqrt{y_1^2 + y_2^2}}$$

$$y_2' = y_1 - \frac{\alpha y_2 y_3}{\sqrt{y_1^2 + y_2^2}}$$

$$y_3' = \frac{\alpha y_1}{\sqrt{y_1^2 + y_2^2}}$$

where α is a parameter. One can easily verify that one trajectory of this system is

$$y_1 = (2 + \cos \alpha t)\cos t,\ y_2 = (2 + \cos \alpha t)\sin t,\ y_3 = \sin \alpha t.$$

EXERCISE

Verify that the above functions y_1, y_2, y_3 solve the system.

This trajectory always lies on the surface of the torus in E^3 obtained by rotating the circle $(y_1 - 2)^2 + y_2^2 = 1$ about the y_3 axis; see Figure 1–8. In order for the trajectory to intersect itself at two distinct times t_1 and t_2, one would need

$$(y_1^2 + y_2^2)^{1/2} - 2 = \cos \alpha t_1 = \cos \alpha t_2, \qquad \frac{y_1}{(y_1^2 + y_2^2)^{1/2}} = \cos t_1 = \cos t_2.$$

Therefore

$$\alpha t_1 - \alpha t_2 = 2k\pi$$

and

$$t_1 - t_2 = 2j\pi,$$

implying $\alpha = k/j$. Thus, if α is irrational, the trajectory continually winds about the torus without ever intersecting itself—indeed a complicated geometry.

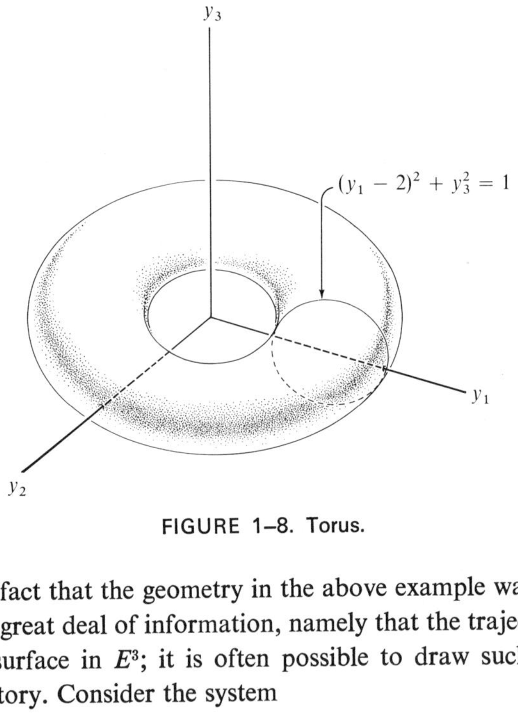

FIGURE 1–8. Torus.

Despite the fact that the geometry in the above example was complicated, we did have a great deal of information, namely that the trajectory remained on a certain surface in E^3; it is often possible to draw such a conclusion about a trajectory. Consider the system

$$\mathbf{y}' = \mathbf{f}(\mathbf{y})$$

in E^n; a surface **S** in E^n is an *integral surface* for this equation if every solution **y** of the differential equation that at some time (t_0) intersects **S** must in fact fall in **S** for all t. Discovery of an integral surface provides geometric insight and also may be of use computationally, as we shall later see.

EXAMPLE Consider the equation $\mathbf{y}' = \mathbf{f}(\mathbf{y})$ where $\mathbf{f}(\mathbf{y})$ defines the vector field in Figure 1–1, that is,

$$y_1' = y_1, \qquad y_2' = y_2.$$

Solving these equations, we find $y_1(t) = y_1(0)e^t$, $y_2(t) = y_2(0)e^t$; therefore, for all solutions, y_1/y_2 is a constant. Since generally one cannot solve complicated systems so easily, we sketch another approach. We seek a function $J(y_1, y_2)$ constant on solutions, that is, so that

$$\frac{\partial J}{\partial y_1} y_1' + \frac{\partial J}{\partial y_2} y_2' = 0.$$

In our case, this asks that

$$\frac{\partial J}{\partial y_1} y_1 + \frac{\partial J}{\partial y_2} y_2 = 0.$$

We seek such a function J. Multiplying by an arbitrary function $a(y_1, y_2)$, we have

$$\frac{\partial J}{\partial y_1} a(y_1, y_2) y_1 + \frac{\partial J}{\partial y_2} a(y_1, y_2) y_2 \equiv 0,$$

and therefore we try

$$\frac{\partial J}{\partial y_1} = a(y_1, y_2) y_2, \qquad \frac{\partial J}{\partial y_2} = -a(y_1, y_2) y_1.$$

For our simple example, we can take $a(y_1, y_2) \equiv a(y_2)$ and discover that $a(y_2) \equiv 1/y_2^2$ allows us to solve easily for J and find $J(y_1, y_2) \equiv y_1/y_2$ as before. ●

EXERCISE

Provide the details in the determination of $a(y_2)$ and $J(y_1, y_2)$ above. Verify that

$$\left(\frac{y_1}{y_2}\right)' = \frac{y_2 y_1' - y_1 y_2'}{y_2^2} = \frac{y_2 y_1 - y_1 y_2}{y_2^2} = 0,$$

so that y_1/y_2 is constant along solutions.

EXERCISE

As in the above example, determine an integral surface for the system $y_1' = 2$, $y_2' = y_1$ and verify the result.

Integral surfaces often represent the mathematical formulation of physical conservation laws. For example, consider the nearly planar motion of the planets about the sun. If the center of the sun is chosen as the origin of a coordinate system in one of these planes, the motion of a planet is approximately described (in appropriate units) by

$$x_1'' = -\frac{x_1}{r^3}$$

$$x_2'' = \frac{x_2}{r^3} \qquad \text{where } r = (x_1^2 + x_2^2)^{1/2}.$$

One integral of the motion is described by the conservation of energy law:

$$\tfrac{1}{2}(x_1'^2 + x_2'^2) - \frac{1}{r} = \text{constant},$$

and another by the conservation of angular momentum law:

$$x_1x_2' - x_2x_1' = \text{constant}.$$

EXERCISE

Verify that the two expressions above are integrals.

Each integral surface represented by one of these equations for a particular value of the constant is a surface in E^4 with coordinates (x_1, x_2, x_1', x_2').

Additional references: 10, 20, 24, 38, 43, 61, 74, 78, 107, 108.

2

INITIAL VALUE PROBLEMS

2.1 INTRODUCTION

From the geometric viewpoint of Chapter 1, it seems reasonable, for well-behaved $\mathbf{f}$, that once we pick any point $\mathbf{y}_0$ in E^n, the differential equation

$$\mathbf{y}' = \mathbf{f}(\mathbf{y}) \tag{2–1}$$

completely determines a trajectory satisfying Eq. 2–1 and passing through the point $\mathbf{y}_0$. Precisely put, the above statement concerns the *initial value problem:* find a function $\mathbf{y}(t)$ such that

$$\mathbf{y}' = \mathbf{f}(\mathbf{y}), \qquad \mathbf{y}(t_0) = \mathbf{y}_0.$$

Without loss of generality we may take $t_0 = 0$. Since the solution $\mathbf{y}(t)$ depends on the initial value $\mathbf{y}_0$, we often will write $\mathbf{y}(t; \mathbf{y}_0)$ to make the dependence explicit. Before proceeding with a study of the general situation, let us look for the moment at the case in which $\mathbf{f}(\mathbf{y})$ is linear and homogeneous in $\mathbf{y}$.

2.2 LINEAR EQUATIONS WITH REAL, CONSTANT COEFFICIENTS

As in Section 1.3, consider the equation

$$\mathbf{y}' = \mathbf{A}\mathbf{y}, \qquad \mathbf{y} \in E^n, \tag{2–2}$$

with initial condition $\mathbf{y}(0) = \mathbf{y}_0$ given, where $\mathbf{A}$ is an n-by-n matrix of real constants. As noted earlier, the general solution of Eq. 2–2 is

$$\mathbf{y}(t) = e^{t\mathbf{A}}\mathbf{c},$$

where c is a fixed vector; thus a particular solution for the initial value problem is

$$\mathbf{y}(t) = e^{t\mathbf{A}}\mathbf{y}_0.$$

The behavior of this solution for large time is determined by the eigenvalues of A. More precisely, let the Jordan canonical form for $\mathbf{A}$ be given by

$$\mathbf{A} = \mathbf{S}^{-1}\mathbf{J}\mathbf{S},$$

where the diagonal of $\mathbf{J}$ consists of the eigenvalues of $\mathbf{A}$, the first super-diagonal of $\mathbf{J}$ contains numbers equal either to one or zero, and the remaining elements of $\mathbf{J}$ are zero. Since

$$e^{t\mathbf{A}} = \sum_{i=0}^{\infty} \frac{(t\mathbf{A})^i}{i!} = \sum_{i=0}^{\infty} \frac{t^i}{i!} (\mathbf{S}^{-1}\mathbf{J}\mathbf{S})^i = \mathbf{S}^{-1}\left[\sum_{i=0}^{\infty} \frac{(t\mathbf{J})^i}{i!}\right] \mathbf{S} = \mathbf{S}^{-1}e^{t\mathbf{J}}\mathbf{S},$$

the behavior of the function

$$\mathbf{y}(t) = e^{t\mathbf{A}}\mathbf{y}_0 = \mathbf{S}^{-1}e^{t\mathbf{J}}\mathbf{S}y_0$$

is governed by that of $e^{t\mathbf{J}}$. It is easily seen that if one of the Jordan blocks $\mathbf{J}_\lambda$ in $\mathbf{J}$ looks like

$$\mathbf{J}_\lambda = \begin{pmatrix} \lambda & 1 & 0 & \cdots & 0 \\ 0 & \lambda & 1 & 0 \cdots & \vdots \\ & & \ddots & \ddots & 0 \\ & & & \ddots & 1 \\ 0 & & & & \lambda \end{pmatrix},$$

where $\mathbf{J}_\lambda$ is k by k, then the corresponding block in $e^{t\mathbf{J}}$ is

$$e^{t\mathbf{J}_\lambda} = \begin{pmatrix} 1 & t & t^2/2 & \cdots & \dfrac{t^{k-1}}{(k-1)!} \\ 0 & 1 & \ddots & \ddots & \vdots \\ \vdots & & \ddots & \ddots & t^2/2 \\ & & & \ddots & t \\ 0 & \cdots & & 0 & 1 \end{pmatrix} e^{\lambda t}.$$

EXERCISE

Verify that e^{tJ_λ} has the above form.

Thus, if all the eigenvalues of A have negative real parts, then all solutions of Eq. 2–2 of the form $e^{tA}\mathbf{c}$ tend exponentially to zero as t increases. If all the eigenvalues of A have nonpositive real parts and the purely imaginary ones λ have a full set of eigenvectors according to the multiplicity of λ, then *all* solutions of the form $e^{tA}\mathbf{c}$ are bounded as t grows. If any eigenvalue has positive real part, then *some* solution becomes unbounded with t.

EXERCISE

Verify the above statements concerning the behavior of solutions for large t. Construct examples of the form

$$y_1' = ay_1 + by_2, \qquad y_2' = cy_1 + dy_2,$$

to illustrate all cases.

In many physical problems, linear as well as nonlinear, we have the case of exponentially decaying solutions. Often the magnitude of the negative real parts of the eigenvalues varies greatly (say by a factor of 1000) in one given problem, meaning that some components of the solution decay very quickly, leaving only the slower transient terms noticeable in the solution. Later we will discuss the numerical difficulties in such "stiff" equations.

Thus far we have exhibited a single solution to Eq. 2–2; is it the *only* one? If there were two solutions $\mathbf{w}$ and $\mathbf{z}$ to Eq. 2–2 with the same initial value $\mathbf{c}$ at $t = t_0$, then

$$\|\mathbf{w}(t) - \mathbf{z}(t)\| = \|[\mathbf{w}(t) - \mathbf{z}(t)] - [\mathbf{w}(t_0) - \mathbf{z}(t_0)]\| = \left\|\int_{t_0}^{t}[\mathbf{w}'(\tau) - \mathbf{z}'(\tau)]d\tau\right\| =$$

$$= \left\|\int_{t_0}^{t}\mathbf{A}(\mathbf{w} - \mathbf{z})d\tau\right\| \leq |t - t_0|\ \|A\|\ \max_{t_0 \leq \tau \leq t} \|\mathbf{w}(\tau) - \mathbf{z}(\tau)\|.$$

Thus, for $t_0 \leq t \leq t_0 + T$, we have

$$\max_{t_0 \leq t \leq t_0+T} \|\mathbf{w} - \mathbf{z}\| \leq T\,\|\mathbf{A}\|\ \max_{t_0 \leq t \leq t_0+T} \|\mathbf{w} - \mathbf{z}\|;$$

therefore, if $T\|\mathbf{A}\| < 1$, we must have $\mathbf{w}(t) = \mathbf{z}(t)$ for $t_0 \leq t \leq t_0 + \text{T}$. This implies that the solution $\mathbf{y}(t) = e^{t\mathbf{A}}\mathbf{y}_0$ is the only solution to Eq. 2–2 with $\mathbf{y}(0) = \mathbf{y}_0$, for $0 \leq t \leq 1/2\|\mathbf{A}\| \equiv T$. When the uniqueness result is applied again, the same function $\mathbf{y}(t)$ is the only solution to Eq. 2–2 with $\mathbf{y}(T) = e^{T\mathbf{A}}\mathbf{y}_0$, for $T \leq t \leq 2T$. Proceeding in this fashion, we deduce that a unique solution to the initial value problem for linear homogeneous systems with constant coefficients always exists for all t.

Additional references: 10, 24, 38, 43, 61, 74.

2.3 NONLINEAR EQUATIONS; EXISTENCE AND UNIQUENESS

Naturally, nonlinear equations are in general more complicated than linear ones; for one thing, existence for all t of a solution to the initial value problem

$$\mathbf{y}' = \mathbf{f}(\mathbf{y}), \qquad \mathbf{y} \in E^n, \tag{2–3}$$

$\mathbf{y}(0) = \mathbf{y}_0$, is not guaranteed; nor, if a solution exists, is its uniqueness guaranteed.

EXERCISE

Show that the equation

$$y' = y^2, \qquad y(1) = -1$$

has solution $y(t) = -t^{-1}$, which does not exist at $t = 0$.

EXERCISE

Show that the differential equation

$$y' = +\sqrt{|y|}, \qquad y(0) = 0$$

has two solutions, $y(t) = t^2/4$ and $y(t) = 0$ for all $t \geq 0$.

It is possible [24, p. 6] to prove that, if $\mathbf{f}(\mathbf{y})$ is continuous in the domain $\|\mathbf{y} - \mathbf{y}_0\| \leq b$ and $M \geq \|\mathbf{f}(\mathbf{y})\|$ for $\|\mathbf{y} - \mathbf{y}_0\| \leq b$, then there exists a solution $\mathbf{y}(t)$ to Eq. 2–3 for $0 \leq t \leq b/M$ and satisfying $\mathbf{y}(0) = \mathbf{y}_0$. Uniqueness can be proved under assumptions which allow one to mimic the proof given in the previous section for linear equations. These same assumptions allow, in fact, for a constructive existence proof that uses a method sometimes used for calculations; we now consider *Picard iteration.*

Suppose that $\mathbf{f}(\mathbf{y})$ satisfies the *Lipschitz condition*

$$\|\mathbf{f}(\mathbf{w}) - \mathbf{f}(\mathbf{z})\| \leq L\|\mathbf{w} - \mathbf{z}\|$$

for $\mathbf{w}$ and $\mathbf{z}$ in the set $R = \{\mathbf{y};\ \|\mathbf{y} - \mathbf{y}_0\| \leq b\}$, and let $M \geq \|\mathbf{f}(\mathbf{y})\|$ for all $\mathbf{y}$ in R. Define

$$\mathbf{y}_0(t) \equiv \mathbf{y}_0, \qquad \mathbf{y}_{i+1}(t) = \mathbf{y}_0 + \int_0^t \mathbf{f}(\mathbf{y}_i(\tau))d\tau \qquad \text{for } i = 0, \ldots$$

For $i = 0$, $\mathbf{y}_i(t)$ is in R for all t, and in particular for $0 \leq |t| \leq T_0 = b/M$. If $\mathbf{y}_i(t)$ is in R for $0 \leq |t| \leq T_0$, then

$$\|\mathbf{y}_{i+1}(t) - \mathbf{y}_0\| \leq |t|M \leq b$$

for $0 \leq |t| \leq T_0$, and hence $\mathbf{y}_{i+1}(t)$ is in R for $0 \leq |t| \leq T_0$. Also,

$$\|\mathbf{y}_{i+1}(t) - \mathbf{y}_i(t)\| = \left\|\int_0^t [\mathbf{f}(\mathbf{y}_i(\tau)) - \mathbf{f}(\mathbf{y}_{i-1}(\tau))]\, d\tau\right\| \leq$$

$$\leq L|t| \max_{0 \leq \tau \leq t} \|\mathbf{y}_i(\tau) - \mathbf{y}_{i-1}(\tau)\|.$$

If we let $T < \min(T_0, 1/L)$, then

$$\max_{0 \leq |t| \leq T} \|\mathbf{y}_{i+1}(t) - \mathbf{y}_i(t)\| \leq LT \max_{0 \leq |t| \leq T} \|\mathbf{y}_i(t) - \mathbf{y}_{i-1}(t)\|$$

and $LT < 1$. It then follows easily that the sequence of functions $\mathbf{y}_i(t)$ converges uniformly to a continuous function $\mathbf{y}(t)$, for $0 \leq |t| \leq T$, satisfying the *integral equation*

$$\mathbf{y}(t) = \mathbf{y}_0 + \int_0^t \mathbf{f}(\mathbf{y}(\tau))d\tau,$$

and hence solving the initial value problem. The argument given in Section 2.2 works here with little alteration to show that $\mathbf{y}$ is the only solution for $0 \leq |t| \leq T$. Thus the technique of reformulating the differential equation as an integral equation leads to useful results; we shall return to this approach later.

EXERCISE

Show that $\mathbf{y}$ is the only solution for $0 \leq |t| \leq T$, as in Section 2.2.

EXAMPLE Consider the equation

$$y' = y^2, \qquad y \in E^1, \qquad y(0) = 1.$$

We find

$$y_0(t) \equiv 1,$$

$$y_1(t) = 1 + \int_0^t 1\,d\tau = 1 + t,$$

$$y_2(t) = 1 + \int_0^t [1+\tau]^2 d\tau = 1 + t + t^2 + \tfrac{1}{3}t^3,$$

$$y_3(t) = 1 + t + t^2 + t^3 + \tfrac{2}{3}t^4 + \tfrac{1}{3}t^5 + \tfrac{1}{9}t^6 + \tfrac{1}{63}t^7.$$

The exact solution is

$$y(t) = \frac{1}{1-t} = \sum_{i=0}^{\infty} t^i. \quad \bullet$$

EXERCISE

Find the first two Picard iterates, starting with $y_0(t) \equiv 1$, for the problem $y' = \sqrt{y}$, $y(0) = 1$.

Additional references: 10, 24, 38, 43, 61, 74.

2.4 REPRESENTATION OF SOLUTIONS; LIE SERIES

Thus far we have been able, under certain strong assumptions, to prove for nonlinear equations the existence and uniqueness (local) properties of linear equations. Perhaps the most convenient property of linear equations with constant coefficients is their solvability in the closed form $e^{t\mathbf{A}}\mathbf{y}_0$; can we do this for nonlinear equations?

Clearly it is not possible to give a closed-form solution for arbitrary nonlinear systems. However, if $\mathbf{f}$ is an analytic function, it is possible to give a *formal* solution by the method of *Lie series;* this has important computational uses.

The Lie-series approach is essentially a perturbation one. We seek to solve Eq. 2–3 with $\mathbf{y}(0) = \mathbf{y}_0$. Suppose we have a solution $\bar{\mathbf{y}}(t)$ to

$$\bar{\mathbf{y}}' = \mathbf{g}(\bar{\mathbf{y}}), \qquad \bar{\mathbf{y}}(0) = \mathbf{y}_0.$$

Define the operators

$$\mathbf{F} = \sum_{i=1}^{n} f_i(\mathbf{y}) \frac{\partial}{\partial y_i},$$

$$\mathbf{G} = \sum_{i=1}^{n} g_i(\mathbf{y}) \frac{\partial}{\partial y_i},$$

and $\mathbf{D} = \mathbf{F} - \mathbf{G}$, where f_i, g_i, y_i are the components of the vectors $\mathbf{f}$, $\mathbf{g}$, $\mathbf{y}$. Then a formal solution to Eq. 2–3 is given [53, 65, 67] by the Lie series:

$$\mathbf{y}(t) = \bar{\mathbf{y}}(t) + \sum_{j=0}^{\infty} \int_0^t \frac{(t-\tau)^j}{j!} [\mathbf{D}\mathbf{F}^j\mathbf{y}]_{\mathbf{y}=\bar{\mathbf{y}}(\tau)} d\tau.$$

In particular, if we take $\mathbf{y}(t) \equiv \mathbf{y}_0$ and $\mathbf{g}(\mathbf{y}) \equiv 0$, this becomes

$$\mathbf{y}(t) = \bar{\mathbf{y}}_0 + \sum_{j=1}^{\infty} \left[\frac{t^j\mathbf{F}^j}{j!}\mathbf{y}\right]_{\mathbf{y}=\bar{\mathbf{y}}_0} \equiv [e^{t\mathbf{F}}\mathbf{y}]_{\mathbf{y}=\bar{\mathbf{y}}_0}.$$

In this final form we see the resemblance to the linear case.

EXERCISE

Show that if $\mathbf{f}(\mathbf{y}) \equiv \mathbf{A}\mathbf{y}$, then the Lie series equals $e^{t\mathbf{A}}\mathbf{y}$.

EXAMPLE Consider the form of the series with $\bar{\mathbf{y}}(t) \equiv \mathbf{y}_0$ and $\mathbf{g}(\mathbf{y}) \equiv 0$ applied to the equation

$$y' = y^2, \qquad y \in E^1, \qquad y(0) = 1,$$

having solution $y(t) = 1/(1-t)$. We now have

$$F = y^2 \frac{d}{dy},$$

so we can see that

$$Fy = y^2, \qquad F^2y = F(Fy) = Fy^2 = 2y^3, \qquad F^3y = F(F^2y) = F(2y^3) = 6y^4,$$

and in general

$$F^jy = j!\; y^{j+1}.$$

Therefore,

$$y(t) = 1 + \sum_{j=1}^{\infty} \frac{t^j}{j!}[j!\; y^{j+1}]_{y=1} = 1 + \sum_{j=1}^{\infty} t^j = \frac{1}{1-t}:$$

the exact solution. ●

EXERCISE

Solve $y' = \sqrt{y}$, $y(0) = 1$, by Lie series.

Additional references: 53, 65.

2.5 DEPENDENCE ON INITIAL DATA

For linear equations, the form of the general solution $\mathbf{y}(t) = e^{t\mathbf{A}}\mathbf{y}_0$ shows that the solution depends very simply, in fact linearly, on the initial data $\mathbf{y}_0$. For nonlinear systems, the solution $\mathbf{y}(t; \mathbf{y}_0)$ depends continuously on $\mathbf{y}_0$ in certain general cases, at least for a small range of values of t. In particular [24, p. 22], suppose $\mathbf{f}(\mathbf{y})$ satisfies a Lipschitz condition in the set $R = \{\mathbf{y};\ \|\mathbf{y} - \mathbf{y}_0\| < b\}$, and that a function $\mathbf{w}(t)$ satisfies $\mathbf{w}' = \mathbf{f}(\mathbf{w})$, $\mathbf{w}(0) = \mathbf{y}_0$, $\mathbf{w}(t) \in R$ for $0 \leq t \leq a$. Then there exists a $\delta > 0$ such that for $(t_0, \mathbf{z}_0) \in U = \{(t, \mathbf{z});\ 0 < t < a, \|\mathbf{z} - \mathbf{w}(t)\| < \delta\}$ there exists a unique solution $\mathbf{z}(t)$ to Eq. 2–3 for $0 \leq t \leq a$ satisfying $\mathbf{z}(t_0) = \mathbf{z}_0$. Moreover, $\mathbf{z}(t) = \mathbf{z}(t; \mathbf{z}_0; t_0)$ is jointly continuous in the variables t, $\mathbf{z}_0$, and t_0 for $(t_0, \mathbf{z}_0) \in U$ and $0 < t < a$. This means that any solution to Eq. 2–3 starting near some point of another solution must exist over essentially the same time interval and remain nearby.

EXERCISE

Study the dependence of the solution on the initial data for the problem $y' = \sqrt{y}$, $y(0) = y_0$.

Having thus indicated the continuity of $\mathbf{y}(t; \mathbf{y}_0)$ in $\mathbf{y}_0$, we naturally ask if $\mathbf{y}$ might not depend more smoothly on $\mathbf{y}_0$, that is, if $\partial\mathbf{y}(t; \mathbf{y}_0)/\partial\mathbf{y}_0$ has meaning. This is important for applications. If the *Jacobian* matrix

$$\mathbf{J}(\mathbf{y}) \equiv \mathbf{f}_y(\mathbf{y}) \equiv \frac{\partial \mathbf{f}}{\partial \mathbf{y}}(\mathbf{y})$$

of partial derivatives of $\mathbf{f}$ is continuous in the region R of the preceding paragraph, then $\partial\mathbf{y}(t; \mathbf{y}_0)/\partial\mathbf{y}_0$ exists; we call this matrix the *connection matrix* $\mathbf{C}(t; \mathbf{y}_0)$. Explicitly,

$$C(t; \mathbf{y}_0)_{ij} = \frac{\partial y_i(t; \mathbf{y}_0)}{\partial y_{0j}}.$$

Intuitively, we see

$$\mathbf{C}(t; \mathbf{y}_0)' = \frac{d}{dt}\mathbf{C}(t; \mathbf{y}_0) = \frac{d}{dt}\frac{\partial \mathbf{y}(t; \mathbf{y}_0)}{\partial \mathbf{y}_0} = \frac{\partial}{\partial \mathbf{y}_0}\frac{d}{dt}\mathbf{y}(t; \mathbf{y}_0) =$$

$$= \frac{\partial}{\partial \mathbf{y}_0}\mathbf{f}(\mathbf{y}(t; \mathbf{y}_0)) = \mathbf{J}(\mathbf{y}(t; \mathbf{y}_0))\frac{\partial \mathbf{y}(t; \mathbf{y}_0)}{\partial \mathbf{y}_0} = \mathbf{J}(\mathbf{y}(t; \mathbf{y}_0))\mathbf{C}(t; \mathbf{y}_0).$$

Since

$$\mathbf{y}(t; \mathbf{y}_0) = \mathbf{y}_0 + t\mathbf{f}(\mathbf{y}_0) + o(t),$$

it follows that

$$\mathbf{C}(0; \mathbf{y}_0) = \mathbf{I}, \qquad \text{the identity matrix.}$$

Thus, $\mathbf{C}(t; \mathbf{y}_0)$ itself satisfies the initial value problem

$$\mathbf{C}(t; \mathbf{y}_0)' = \mathbf{J}(\mathbf{y}(t; \mathbf{y}_0))\mathbf{C}(t; \mathbf{y}_0), \qquad \mathbf{C}(0; \mathbf{y}_0) = \mathbf{I},$$

a *linear* differential equation for $\mathbf{C}(t; \mathbf{y}_0)$. A rigorous argument of this type can be used to prove [24, p. 25] that $\mathbf{C}(t; \mathbf{y}_0)$ exists and that the determinant $\det \mathbf{C}(t; \mathbf{y}_0)$ satisfies

$$\det \mathbf{C}(t; \mathbf{y}_0) = \exp\{\int_0^t \operatorname{tr} \mathbf{J}(\mathbf{y}(s; \mathbf{y}_0))ds\},$$

where $\operatorname{tr} \mathbf{A}$ for a matrix $\mathbf{A}$ is $\operatorname{tr} \mathbf{A} = \sum_{i=1}^n A_{ii}$, the trace of $\mathbf{A}$; therefore $\mathbf{C}(t; \mathbf{y}_0)$ is not singular. We will be able to make computational use of $\mathbf{C}(t; \mathbf{y}_0)$, computing it from its differential equation in later sections.

EXERCISE

Find the connection matrix (scalar) for $y' = \sqrt{y}$, $y(0) = y_0 > 0$.

Having just indicated that in many cases a small change in initial data induces a small change throughout the solution on a bounded interval, we give a trivial example to restrain the overly enthusiastic reader. Consider the system

$$y_1' = 1$$

$$y_2' = y_2 + \cos y_1 - \sin y_1.$$

With initial conditions $y_1 = y_2 = 0$, the solution is

$$y_1(t) = t, \qquad y_2(t) = \sin t.$$

With $y_1(0) = 0$, $y_2(0) = \epsilon \neq 0$, however, the solution is

$$y_1(t) = t, \qquad y_2(t) = \epsilon e^t + \sin t;$$

although on any bounded interval this is a small change for any small ϵ, as the theory indicates, the change is certainly significant for large t. Later we will see that often one can guarantee that the change remains small for all $t \geq 0$.

Additional references: 10, 24, 38, 43, 61, 74, 78.

2.6 PERTURBATIONS AND EXPANSIONS

It is reasonable to ask whether or not the continuity of the solution, in its dependence on the initial data, extends in some sense to continuity with respect to changes in the equation; that is, do nearby equations have nearby solutions? This opens the vast subject of perturbations and stability; in this and the next section we look very briefly at such concepts.

Perhaps the oldest and simplest perturbation concept is as follows. Suppose we can solve easily

$$\mathbf{y}' = \mathbf{f}(\mathbf{y}), \qquad \mathbf{y}(0) = \mathbf{y}_0,$$

and wish now to solve

$$\mathbf{z}' = \mathbf{f}(\mathbf{z}) + \epsilon\mathbf{g}(\mathbf{z}), \qquad \mathbf{z}(0) = \mathbf{z}_0$$

for small ϵ. We try

$$\mathbf{z}(t) = \mathbf{u}_0(t) + \epsilon\mathbf{u}_1(t) + \ldots,$$

getting

$$\begin{aligned} \mathbf{u}_0' + \epsilon\mathbf{u}_1' + \ldots &= \mathbf{f}(\mathbf{u}_0 + \epsilon\mathbf{u}_1 + \ldots) + \epsilon\mathbf{g}(\mathbf{u}_0 + \epsilon\mathbf{u}_1 + \ldots) \\ &= \mathbf{f}(\mathbf{u}_0) + \epsilon\mathbf{J}(\mathbf{u}_0)\mathbf{u}_1 + \epsilon\mathbf{g}(\mathbf{u}_0) + O(\epsilon^2), \end{aligned}$$

where $\mathbf{J}(\mathbf{u}_0)$ is the Jacobian matrix of partial derivatives of $\mathbf{f}$ evaluated at $\mathbf{u}_0$. Matching terms in ϵ, we find

$$\mathbf{u}_0' = \mathbf{f}(\mathbf{u}_0), \qquad \mathbf{u}_0(0) = \mathbf{z}_0, \qquad \mathbf{u}_1' = \mathbf{J}(\mathbf{u}_0)\mathbf{u}_1 + \mathbf{g}(\mathbf{u}_0), \qquad \mathbf{u}_1(0) = \mathbf{0}$$

et cetera, giving us the terms in a perturbation expansion for the solution $\mathbf{u}(t)$. Much of our knowledge about the motions of the planets and their satellites in our solar system has been derived in this way via perturbations of previous results, that is, by starting with the two-body model (sun and planet) and one by one adding perturbing forces caused by other planets, oblateness, et cetera.

EXAMPLE Consider the equation for a nonlinear spring,

$$y'' + y + y^3 = 0, \qquad y(0) = \alpha, \qquad y'(0) = 0,$$

where we assume that α is small. Letting

$$y = \alpha x(t),$$

we find

$$x'' + x + \alpha^2 x^3 = 0, \qquad x(0) = 1, \qquad x'(0) = 0.$$

Applying the expansion

$$x(t) = u_0(t) + \alpha^2 u_1(t) + \ldots,$$

we find

$$u_0'' + u_0 = 0, \qquad u_0(0) = 1, \qquad u_0'(0) = 0,$$

$$u_1'' + u_1 + u_0^3 = 0, \qquad u_1(0) = u_1'(0) = 0,$$

et cetera. This yields finally

$$y(t) = \alpha\left\{\cos t + \alpha^2\left[\frac{\cos 3t}{32} - \frac{\cos t}{32} - \frac{3t \sin t}{8}\right]\right\} + O(\alpha^5). \quad \bullet$$

EXERCISE

Fill in the missing details in the example above.

The validity of the perturbation expansion can, for many classes of equations, be demonstrated rigorously. We warn the reader, however, to be careful about such *singular perturbation* problems as

$$\epsilon u'' + u' + u = 0, \qquad u(0) = 1, \qquad u'(0) = 0,$$

for which the perturbation expansion yields nonsense; such problems where the small coefficient appears with the highest derivative require special techniques that are beyond the scope of this book. See, for example, J. Cole's *Perturbation methods in applied mathematics* (Blaisdell, 1969).

Returning to the simple idea that nearby equations should have nearby solutions, we give a brief indication of the validity of this concept in a special but important situation. Suppose that all solutions of the linear system

$$\mathbf{y}' = \mathbf{A}\mathbf{y}$$

converge to zero as t tends to infinity, that the initial vector $\mathbf{z}_0$ is sufficiently small, and that the continuous function

$$\mathbf{g}(t, \mathbf{z}) = o(\|\mathbf{z}\|) \quad \text{as} \quad \|\mathbf{z}\| \to 0$$

uniformly for $t \geq 0$; then [24, p. 314–317] the solution of

$$\mathbf{z}' = \mathbf{A}\mathbf{z} + \mathbf{g}(t, \mathbf{z}), \qquad \mathbf{z}(0) = \mathbf{z}_0 \tag{2–4}$$

converges to zero as t tends to infinity, and $\sup_{t \geq 0}\|\mathbf{z}(t)\|$ tends to zero with $\|\mathbf{z}_0\|$. If instead some eigenvalue of $\mathbf{A}$ has positive real part then $\sup_{t \geq 0}\|\mathbf{z}(t)\|$ does not converge to zero with $\|\mathbf{z}_0\|$. Briefly, we say that the zero solution of Eq. 2–4 is *asymptotically stable* if all eigenvalues of $\mathbf{A}$ have negative real parts, but is *not stable* if any eigenvalue of $\mathbf{A}$ has positive real part. It is possible to give much more precise statements about the behavior of the solutions, both in the small and in the large. For clarity we look at results in the small for two-dimensional systems before we proceed to asymptotic results in general.

EXERCISE

Study the stability of the zero solution of

$$z_1' = az_1 + bz_2 + z_2^2 \sin t$$

$$z_2' = z_1 + az_2 + z_2^2 \cos t$$

in the cases: (i) $a < 0$, $b \leq 0$; (ii) $a < 0$, $0 < b < a^2$; (iii) $a < 0$, $b > a^2$; (iv) $a > 0$.

Additional references: 6, 7, 18, 24.

2.7 PERTURBATIONS OF TWO-DIMENSIONAL LINEAR SYSTEMS

Consider the linear system $\mathbf{y}' = \mathbf{A}\mathbf{y}$ in two dimensions, that is

$$\begin{aligned} y_1' &= ay_1 + by_2 \\ y_2' &= cy_1 + dy_2 \end{aligned} \tag{2–5}$$

with $ad - bc \neq 0$. The behavior of solutions to Eq. 2–5 is determined, as we saw in Section 2.2, by the eigenvalues λ and μ of **A**; either λ and μ are real or, if $\lambda = \alpha + i\beta$ and $\beta \neq 0$, then $\mu = \alpha - i\beta$. By a change of coordinates, the matrix **A** can be reduced to one of several forms with corresponding types of trajectories, as given by the various cases in Figure 2–1.

Consider the perturbed system

$$\begin{aligned} y_1' &= ay_1 + by_2 + g_1(y_1, y_2) \\ y_2' &= cy_1 + dy_2 + g_2(y_1, y_2), \end{aligned} \tag{2–6}$$

where g_1 and g_2 are continuous and satisfy, say,

$$|g_i(\mathbf{y})| = O(\|\mathbf{y}\|^{1+\epsilon}),\ \epsilon > 0, \text{ as } \|\mathbf{y}\| \to 0,\ i = 1, 2.$$

Then one can show [24, Ch. 15]: (I) if (0, 0) is a *spiral point* (Case vii or ix of Figure 2–1) for Eq. 2–5, it is also for Eq. 2–6; (II) if (0, 0) is a *proper node* (Case i or ii) for Eq. 2–5, it is also for Eq. 2–6; (III) if (0, 0) is a *center* (Case x or xi) for Eq. 2–5, it is either a center or a spiral point for Eq. 2–6; (IV) if (0, 0) is an *improper node* (Case iii) for Eq. 2–5, then every solution of Eq. 2–6 starting near (0, 0) tends to (0, 0) with a limiting direction at an angle of 0, $\pi/2$, π, or $3\pi/2$ with the positive y_1 axis; (V) if (0, 0) is a *saddle point* (Case vii) for Eq. 2–5, there exists at least one solution of Eq. 2–6 tending to (0, 0) at each of the angles 0 and π; (VI) if (0, 0) is an *attractor* (Case i, iii, v, or vii) for Eq. 2–5, it is also for Eq. 2–6; (VII) if (0, 0) is a *repeller* (Case ii, iv, vi, viii, or ix) for Eq. 2–5, it is also for Eq. 2–6.

These results above allow one to analyze the behavior of solutions near points where the system looks like Eq. 2–6 with $\mathbf{g}(\mathbf{y})$ as described.

EXAMPLE Consider the system

$$\begin{aligned} y_1' &= y_2 \\ y_2' &= -\sin y_1 \end{aligned} \tag{2–7}$$

which has *rest points,* where $y_1' = y_2' = 0$, on the line $y_2 = 0$ at $y_1 = m\pi$, $m = 0, \pm 1, \pm 2, \ldots$. Near these points the system has the form

$$\begin{aligned} y_1' &= y_2 \\ y_2' &= -(\cos m\pi)(y_1 - m\pi) + O(y_1 - m\pi)^2, \end{aligned}$$

so that near $y_1 = 0, \pm 2\pi, \pm 4\pi, \ldots$ the system is similar to that linear system with matrix

$$\begin{pmatrix} 0 & 1 \\ -1 & 0 \end{pmatrix},$$

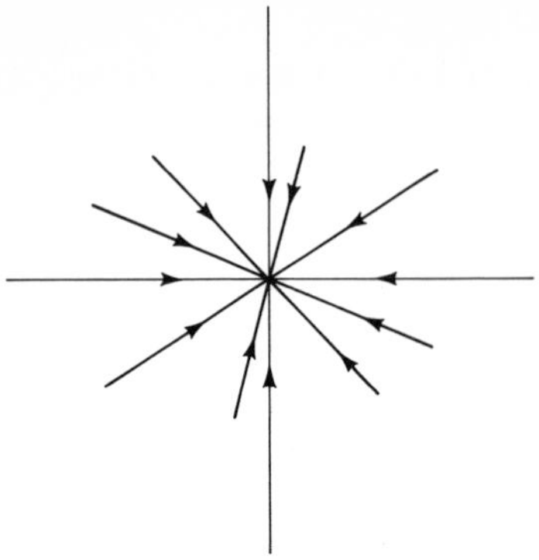

i) $A = \begin{pmatrix} \lambda & 0 \\ 0 & \lambda \end{pmatrix}, \lambda < 0$

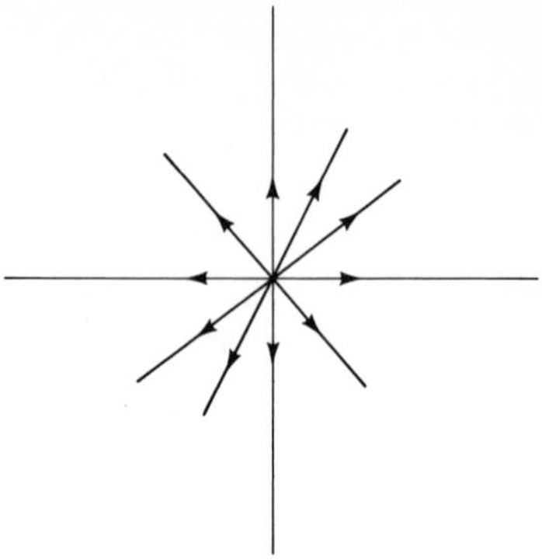

ii) $A = \begin{pmatrix} \lambda & 0 \\ 0 & \lambda \end{pmatrix}, \lambda > 0$

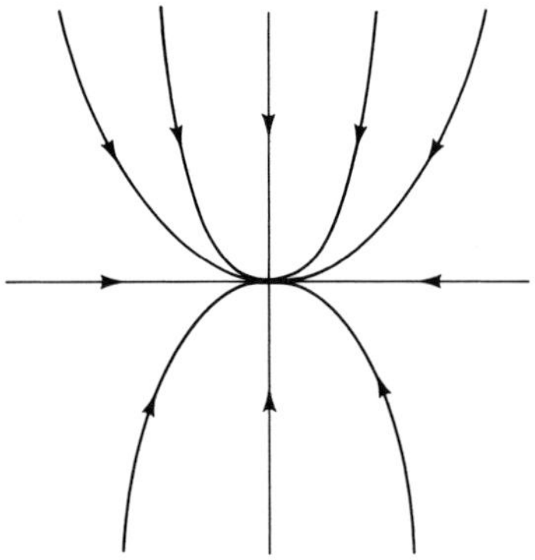

iii) $A = \begin{pmatrix} \lambda & 0 \\ 0 & \mu \end{pmatrix}, \mu < \lambda < 0$

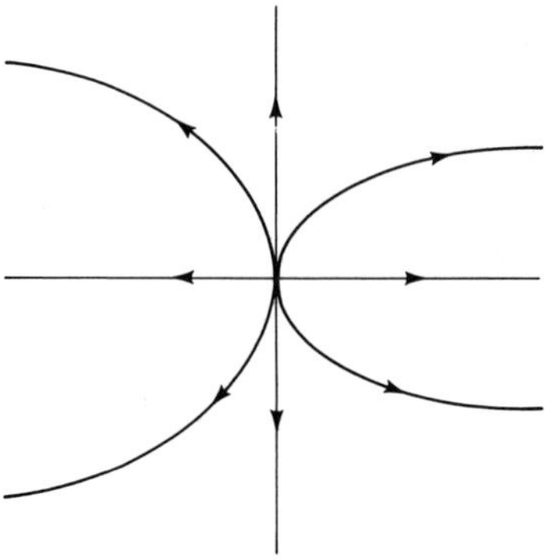

iv) $A = \begin{pmatrix} \lambda & 0 \\ 0 & \mu \end{pmatrix}, 0 < \mu < \lambda$

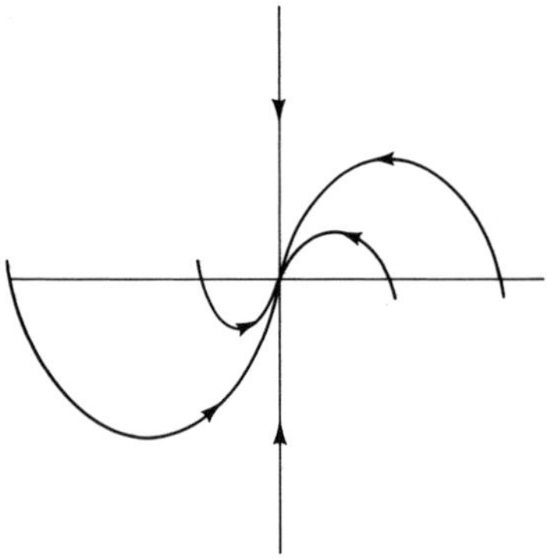

v) $A = \begin{pmatrix} \lambda & 0 \\ 1 & \lambda \end{pmatrix}, \lambda < 0$

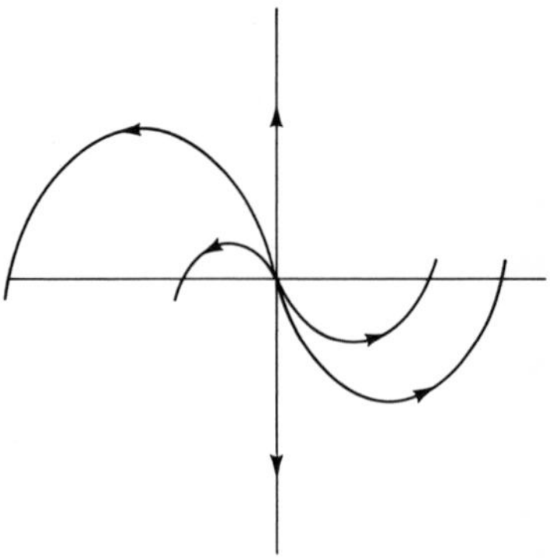

vi) $A = \begin{pmatrix} \lambda & 0 \\ 1 & \lambda \end{pmatrix}, \lambda > 0$

FIGURE 2–1, i-vi

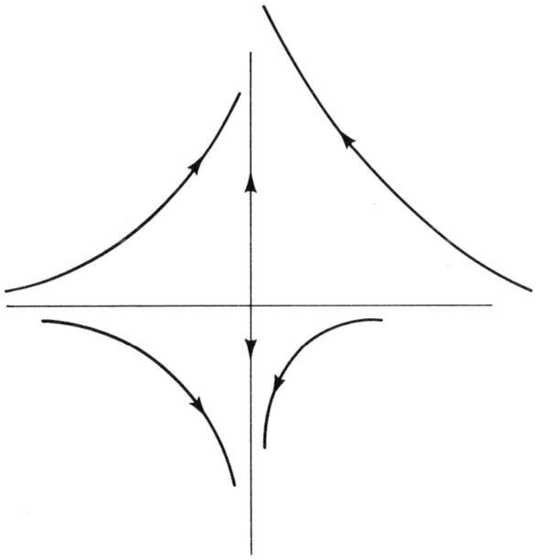

vii) $A = \begin{pmatrix} \lambda & 0 \\ 0 & \mu \end{pmatrix}, \lambda < 0 < \mu$

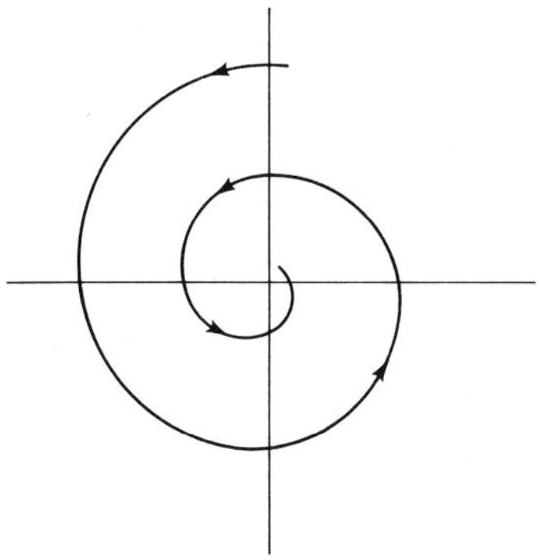

viii) $A = \begin{pmatrix} \alpha & \beta \\ -\beta & \alpha \end{pmatrix}, \alpha < 0, \beta < 0$

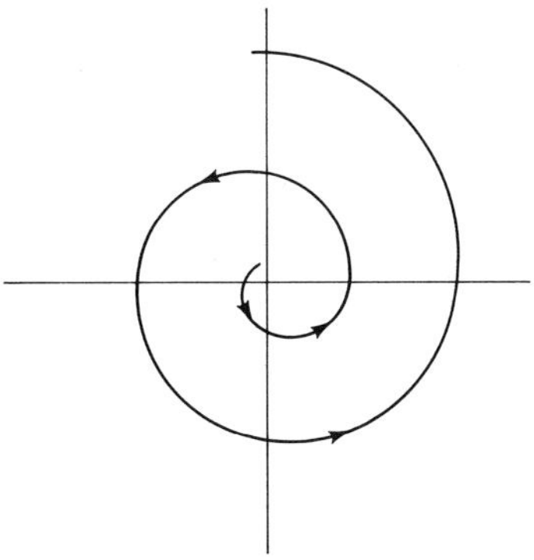

ix) $A = \begin{pmatrix} \alpha & \beta \\ -\beta & \alpha \end{pmatrix}, \alpha > 0, \beta < 0$

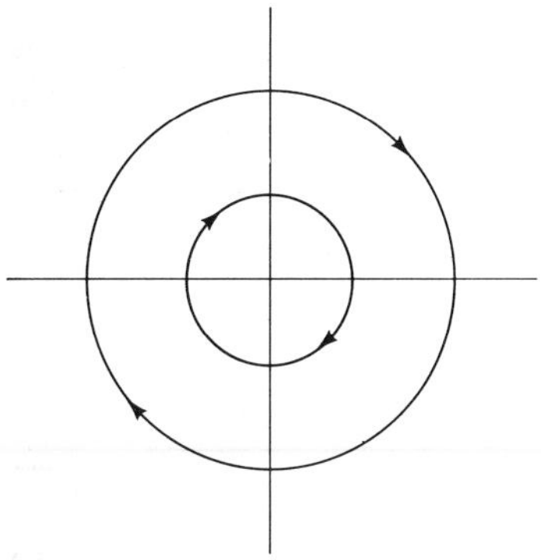

xi) $A = \begin{pmatrix} 0 & \beta \\ -\beta & 0 \end{pmatrix}, \beta > 0$

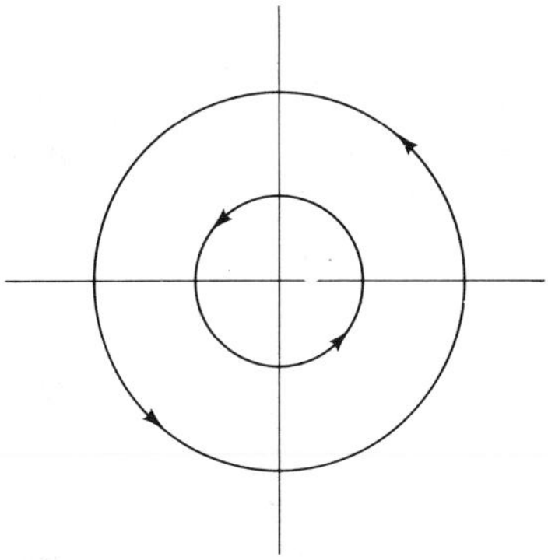

x) $A = \begin{pmatrix} 0 & \beta \\ -\beta & 0 \end{pmatrix}, \beta < 0$

FIGURE 2–1, vii-xi

while the relevant matrix near $y_1 = \pm\pi, \pm 3\pi, \ldots$ is

$$\begin{pmatrix} 0 & 1 \\ 1 & 0 \end{pmatrix};$$

the eigenvalues are $\pm i$ in the first place and ± 1 in the second, so we have the Cases xi and vii. Putting these pictures together we have the following partial picture of the solution curves for Eq. 2–7. ●

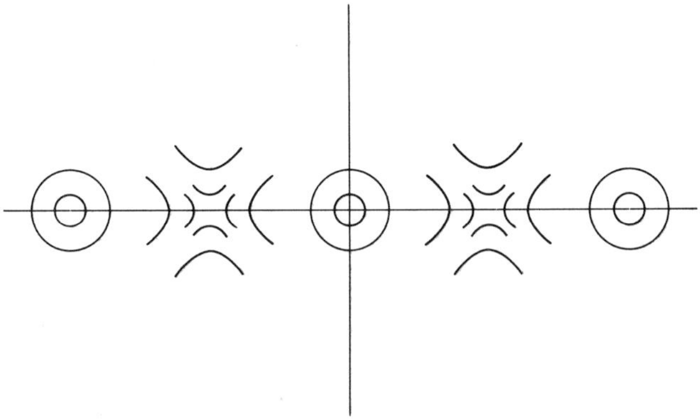

FIGURE 2–2

EXERCISE

Fill in details in the above example.

Using the fact that $-\cos y_1 + y_2^2/2$ is an integral of the system Eq. 2–7, we can see that the perturbation method gives a good idea of the solutions, which actually look qualitatively like Figure 2–3.

EXERCISE

Verify that $-\cos y_1 + y_2^2/2$ is an integral of the system.

This approach can of course be used near rest points of any system to linearize the equations and obtain a qualitative picture of the trajectories near that rest point. If we consider the system

$$\mathbf{y}' = \mathbf{f}(\mathbf{y}) = \mathbf{f}(\mathbf{y}_0) + \mathbf{J}(\mathbf{y}_0)(\mathbf{y} - \mathbf{y}_0) + o(\|\mathbf{y} - \mathbf{y}_0\|)$$

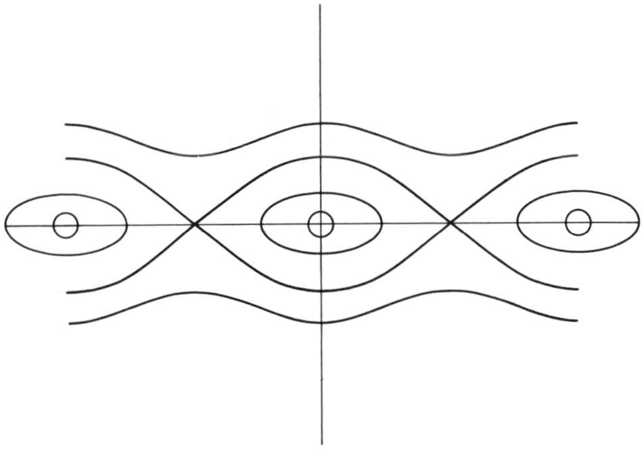

FIGURE 2–3

near the rest point $\mathbf{y}_0$ where $\mathbf{f}(\mathbf{y}_0) = 0$, we can think of $\mathbf{y}(t)$ as a perturbation of $\mathbf{y}_0 + \mathbf{z}(t)$ where $\mathbf{z}$ solves

$$\mathbf{z}' = \mathbf{J}(\mathbf{y}_0)\mathbf{z}.$$

Since the equation for $\mathbf{y}(t)$ near $\mathbf{y}_0$ is a small perturbation of that for $\mathbf{z}(t)$, the solutions of the linearized equation for $\mathbf{z}(t)$ will describe rather well the behavior of the solutions $\mathbf{y}(t)$ near $\mathbf{y}_0$.

Additional reference: 24.

2.8 AN APPLICATION TO THE WEATHER

We know from the study of weather reports that the distribution of atmospheric pressure has something to do with wind velocities; high-pressure areas are generally associated with a calm and low-pressure areas with winds and turbulance. Given a nonconstant pressure distribution p over the earth's surface, which we assume is the sphere $S_2 = \{(y_1, y_2, y_3); y_1^2 + y_2^2 + y_3^2 = 1\}$ in E^3 with appropriate units, consider a small patch of the surface and two neighboring isobars (curves on which p is constant) with values p_1 and p_2. If the isobars are differentiable curves, then we can describe mathematically the force on the strip of air between the isobars produced by the pressure difference $p_2 - p_1$. Let $\mathbf{T}(\mathbf{y})$ be the vector which is normal to the curve $p(\mathbf{y}) =$ constant and lies in the tangent plane to S_2 at $\mathbf{y}$; let $\partial p / \partial \mathbf{T}$ denote the vector with direction $\mathbf{T}(\mathbf{y})$ and magnitude equal to the directional derivative of p in the direction $\mathbf{T}$ at $\mathbf{y}$. We then have a force field on S_2 given by

$\partial p(\mathbf{y})/\partial \mathbf{T}$, and the second-order differential equation (Newton's second law of motion)

$$\frac{d^2\mathbf{y}}{dt^2} \equiv \mathbf{y}'' = \frac{-\partial p(\mathbf{y})}{\partial \mathbf{T}}, \qquad \mathbf{y} \in S_2 \subset E^3$$

describes a law of motion which we will assume for a "wind particle" moving on the surface S_2.

We suppose further that $\partial p(\mathbf{y})/\partial \mathbf{T}$ is a continuous, time-independent vector field on S_2. The assumption of time independence gives only a rough approximation to the facts, of course; however, for short intervals of time it may lead to reasonable predictions of wind velocity since the observed distortion of a pattern of isobars is slower than wind velocities and we can refer our coordinate systems to a "standing" pattern of isobars. A storm center, in particular (usually associated with a low-pressure center), moves more slowly than the winds in the storm.

It is known that any continuous vector field on S_2 must vanish at some point in S_2.* Let $\mathbf{y}_0$ be that point. Thus, we can suppose that at $\mathbf{y}_0$ and $t = 0$ we have $\mathbf{y}' = 0$ and $\partial p/\partial \mathbf{T} = 0$.

If the isobars look like ellipses near $\mathbf{y}_0$, where $\partial p/\partial \mathbf{T} = 0$, say

$$p(\mathbf{y}) = p_0 + \tfrac{1}{2}(ay_1^2 + by_2^2)$$

in a local coordinate patch at $\mathbf{y}_0$, then linearization about the rest point yields integral curves in the y_1', y_2' plane, approximately satisfying the system

$$y_2'' + ay_1 = 0$$

$$y_2'' + by_2 = 0$$

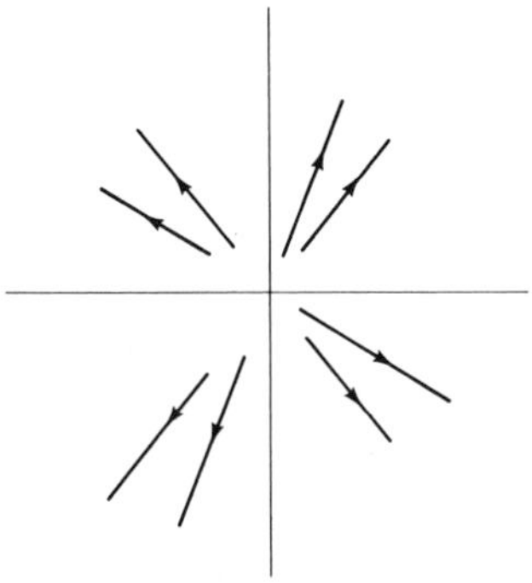

FIGURE 2–4

*Since the proof of this is rather deep, we merely suggest that, as an exercise, the reader attempt to draw a continuous nonvanishing vector field of arrows on a basketball. For a proof, see Berger and Berger, *Perspectives in nonlinearity*, New York, Benjamin, 1968.

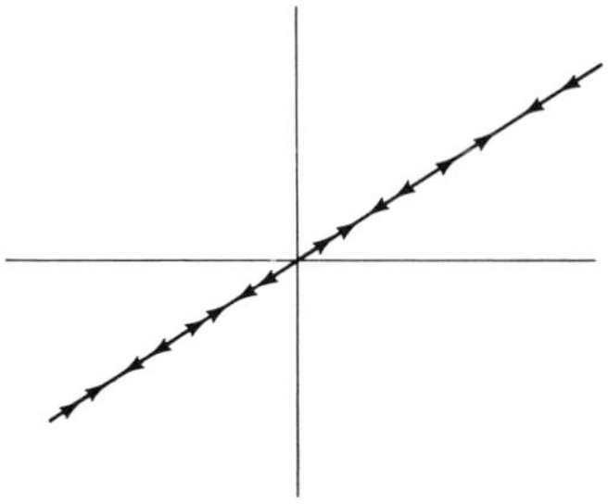

FIGURE 2–5

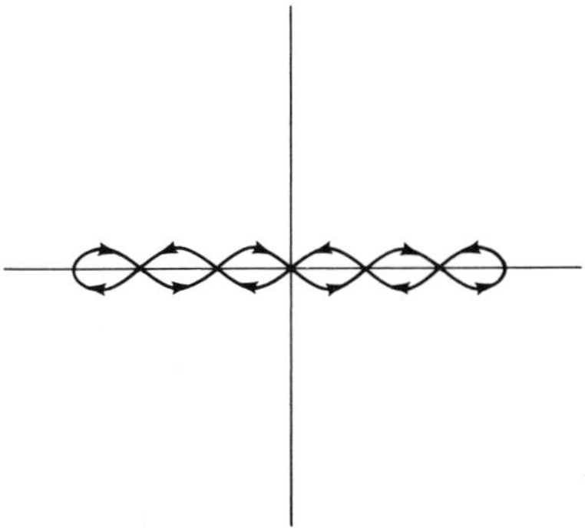

FIGURE 2–6

and looking like Figure 2–4 for a high-pressure center at $\mathbf{y}_0$ with

$$y_1' \approx -\sqrt{|a|}\ \sinh \sqrt{|a|}\,t$$
$$y_2' \approx -\sqrt{|b|}\ \sinh \sqrt{|b|}\,t$$
$$a < 0, \qquad b < 0,$$

and like Figure 2–5 (for $a = b > 0$) or Figure 2–6 with

$$y_1' \approx \sqrt{a} \sin \sqrt{a}\ t$$
$$y_2' \approx \sqrt{b} \sin \sqrt{b}\ t,$$
$$a > b > 0.$$

According to the model of atmospheric motion adopted here, there will always be at least one point near which the wind motion is of one of the above types or, in case the isobars near the rest point $\mathbf{y}_0$ are hyperbolic curves, e.g., $a > 0$, $b < 0$ (a saddle point $\mathbf{y}_0$ of $p(\mathbf{y})$), the wind motions will

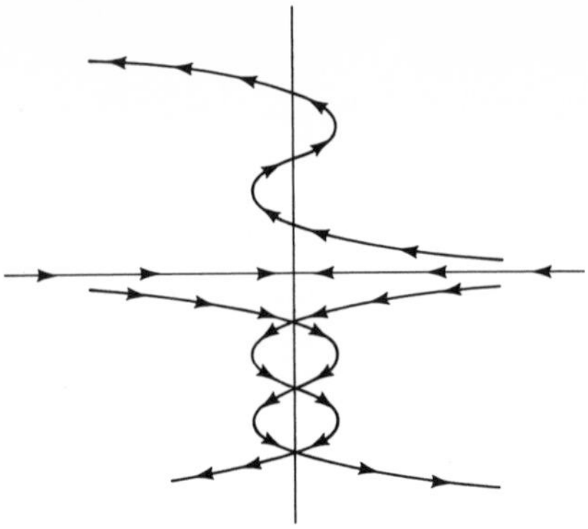

FIGURE 2–7

look like those in Figure 2–7 with

$$y_1' \approx -\sqrt{a}\sin\sqrt{a}\,t$$

$$y_2' \approx -\sqrt{|b|}\,\sinh\sqrt{|b|}\,t.$$

According to our model, there will be at least this difference between wind flows over a sphere and those over an infinite plane: (1) over an infinite plane E^2, it is possible to have uniform translational flow and things topologically equivalent to it (smooth "laminar" flows); (2) over a 2-sphere S_2, there will be at least one point near which the flow will be turbulent. Thus, the variations in winds from place to place and time to time are due, at least in part, to the topological character of the earth's surface—the fact that it is essentially a 2-sphere.

EXERCISE

Fill in the details in the above presentation.

EXERCISE

Examine the "parabolic" case $a = 0$, $b \neq 0$ for the above presentation.

2.9 ASYMPTOTIC BEHAVIOR

Having seen at the end of Section 2.7 how one can examine behavior in the small near rest points by using linearization, we proceed to investigate behavior for large time to get asymptotic results.

In Section 2.6 we stated the result that, for small enough $\|\mathbf{z}_0\|$, all solutions of Eq. 2–4 converge to zero as t tends to infinity if the eigenvalues of $\mathbf{A}$ have negative real parts and if $\mathbf{g}(t, \mathbf{z}) = o(\|\mathbf{z}\|)$; this is perhaps the most well-known result on asymptotic behavior of solutions. A somewhat similar but more general result is the following: Suppose that the eigenvalues of $\mathbf{A}$ have negative real parts, that $\mathbf{g}(t, \mathbf{y})$ tends to zero as t tends to infinity, uniformly in $\mathbf{y}$ for small enough $\|\mathbf{y}\|$, and that for any $\epsilon > 0$ there exists $\delta_\epsilon > 0$ and T_ϵ such that $\|\mathbf{f}(t, \mathbf{y})\| \leq \epsilon\|\mathbf{y}\|$ for $\|\mathbf{y}\| < \delta_\epsilon$ and $t \geq T_\epsilon$. Then [24, p. 327] there exists a T such that any solution $\mathbf{y}(t)$ of

$$\mathbf{y}' = \mathbf{A}\mathbf{y} + \mathbf{f}(t, \mathbf{y}) + \mathbf{g}(t, \mathbf{y}), \qquad \mathbf{f} \text{ and } \mathbf{g} \text{ continuous},$$

converges to zero as t tends to infinity, if $\|\mathbf{y}(T)\|$ is small enough. The value of T can be computed as follows: Choose K and σ such that $\|e^{t\mathbf{A}}\| \leq Ke^{-t\sigma}$, let $\epsilon < \sigma/2K$, and choose δ_ϵ and T_ϵ accordingly. Select T' such that $\|\mathbf{g}(t, \mathbf{y}\| \leq a < (\sigma - K_\epsilon/K)\delta_\epsilon$ for $t \geq T'$ and $\|\mathbf{y}\| \leq \delta_\epsilon$; then $T = \max(T_\epsilon, T')$. Moreover, one can show that "$\|\mathbf{y}(T)\|$ small enough" means

$$\|\mathbf{y}(T)\| \leq \frac{\delta_\epsilon(\sigma - K_\epsilon)}{K\sigma} - \frac{a}{\sigma}.$$

EXAMPLE Consider the system

$$y_1' = -2y_1 + y_2^2 + e^{-t} \cos y_2$$

$$y_2' = -y_2 + y_1^2 \cos t + e^{-t^2} y_1.$$

Here we have

$$\mathbf{A} = \begin{pmatrix} -2 & 0 \\ 0 & -1 \end{pmatrix},$$

so, using the vector norm $\|(y_1, y_2)\|_\infty = \max(|y_1|, |y_2|)$, we have $\|e^{t\mathbf{A}}\|_\infty \leq 1 \cdot e^{-t}$, and $K = \sigma = 1$. Thus we need $\epsilon < \frac{1}{2}$, say $\epsilon = \frac{1}{3}$. Now $\mathbf{f}(t, \mathbf{y}) = (y_2^2, y_1^2 \cos t)$, so $\|\mathbf{f}(t, \mathbf{y})\|_\infty \leq \|\mathbf{y}\|_\infty^2$ for all t; thus we can take $\delta_\epsilon = \epsilon = \frac{1}{3}$, $T_\epsilon = 0$. We need $a < (\sigma - K_\epsilon/K) \cdot \delta_\epsilon = (1 - 1 \cdot \frac{1}{3}/1) \cdot \frac{1}{3} = \frac{2}{9}$, say $a = \frac{1}{5}$. $\|\mathbf{g}(t, \mathbf{y})\|_\infty = \|(e^{-t} \cos y_2, e^{-t^2} y_1)\|_\infty \leq e^{-t}\|(\cos y_2, y_1)\|_\infty \leq e^{-t}$ for $\|\mathbf{y}\|_\infty \leq \delta_\epsilon$; thus we may take $T = T' \geq \ln 5$, say $T = 1.7$. Therefore any solution tends to zero if $\|\mathbf{y}(1.7)\|_\infty \leq .02 \leq (1/45)$. ●

In the more general case where only k eigenvalues of $\mathbf{A}$ have negative real parts while n–k have positive real parts, one can show [24, p. 330] that for large t_0 there is a k-parameter "surface" in $\mathbf{y}$-space such that solutions starting there tend to zero while solutions starting near the origin but not on the "surface" do not tend to zero.

A valuable general method for studying asymptotic behavior is due to Lyapanov [18]. A real-valued function $V(\mathbf{y})$ on E^n is *positive* (*negative*) *definite* if there exists a number B such that, for $\|\mathbf{y}\| \leq B$ and $y \neq 0$, $V(\mathbf{y}) > 0$ ($V(\mathbf{y}) < 0$). Consider the system

$$\mathbf{y}' = \mathbf{f}(\mathbf{y}), \tag{2–8}$$

where we assume $\mathbf{f}(\mathbf{0}) = \mathbf{0}$. Suppose there exists a function $V(\mathbf{y})$ on E^n such that (1) $V(\mathbf{y})$ is definite, (2) $\dot{V}(\mathbf{y}) \equiv \sum_{i=1}^{n} (\partial V/\partial y_i) f_i(\mathbf{y})$ is definite, (3) V and $\dot{V}$ have opposite signs, and (4) for every $\epsilon > 0$ there exists $\delta > 0$ such that $\|\mathbf{y}\| < \delta$ implies $|V(\mathbf{y})| < \epsilon$; then any solution of Eq. 2–8 tends to zero if $\|\mathbf{y}(0)\|$ is small enough. Finding an appropriate V is usually difficult; widely applicable techniques are not known.

EXAMPLE Consider the system

$$y_1' = -2y_1 - 2y_1y_2^2$$

$$y_2' = -4y_2 - 2y_2y_1^2.$$

Let $V(\mathbf{y}) = -(y_1^2 + 2y_2^2 + y_1^2y_2^2)$, so that $\dot{V} = (2y_1 + 2y_1y_2^2)^2 + (4y_2 + 2y_2y_1^2)^2$. Clearly $V(\mathbf{y})$ is negative definite and satisfies condition (4) while $\dot{V}$ is positive definite. Therefore, any solution tends to zero if $\|\mathbf{y}(0)\|$ is small enough. In fact, if $\mathbf{y}(t) \neq 0$ is *any* solution, then $d/(dt)\ V(\mathbf{y}) = \dot{V}(\mathbf{y}) > 0$ and therefore $V(\mathbf{y}(t)) < 0$ is increasing with t and must converge to something. It follows that $d/(dt)\ V(\mathbf{y})$ tends to zero and hence, from the nature of $\dot{V}(\mathbf{y})$, *every* solution converges to zero. ●

Additional references: 7, 10, 18, 24, 38, 43, 74.

2.10 PERIODIC SOLUTIONS

A function $\mathbf{y}(t)$ is *periodic with period T* if $\mathbf{y}(t + T) = \mathbf{y}(t)$ for all t. Although periodic solutions to differential equations are of great interest, their existence is usually very difficult to assert. If we consider the linear system

$$\mathbf{y}' = \mathbf{A}\mathbf{y}$$

with general solution

$$\mathbf{y}(t) = e^{t\mathbf{A}}\mathbf{y}(0),$$

it is clear that there exist periodic solutions other than the zero solution if and only if some eigenvalue of $\mathbf{A}$ has its real part equal to zero, that is, is purely imaginary. For *nonlinear* systems in two dimensions, the following is known [24, p. 391]: If any trajectory $\mathbf{y}(t)$ of

$$\mathbf{y}' = \mathbf{f}(\mathbf{y})$$

remains in a closed, bounded subset of E^2, and if the *limit set*

$$\mathbf{L} \equiv \{\mathbf{z};\ \text{for some sequence } t_n \to \infty,\ \|\mathbf{z} - \mathbf{y}(t_n)\| \to 0\}$$

is such that $\mathbf{f}(\mathbf{z}) \neq 0$ for $\mathbf{z}$ in L, then the set L is itself a periodic trajectory for the given system. With the exception of that result, most theorems about periodic solutions are perturbation results.

For example, consider the system

$$\mathbf{y}' = \mathbf{f}(\mathbf{y};\ \epsilon) \tag{2–9}$$

where ϵ is a parameter; suppose that for $\epsilon = 0$ Eq. 2–9 has a real periodic solution $\mathbf{p}(t)$ of period T_0. Suppose $\mathbf{f}(\mathbf{y};\ \epsilon)$ is continuously differentiable in $(\mathbf{y},\ \epsilon)$ for $|\epsilon|$ small in a neighborhood of $\mathbf{p}(t)$; then clearly the function $\mathbf{z}(t) \equiv \mathbf{p}'(t)$ satisfies

$$\mathbf{z}' = \left[\frac{\partial \mathbf{f}}{\partial \mathbf{y}}\right]_{\mathbf{y} = \mathbf{p}(t),\ \epsilon = 0} \mathbf{z}. \tag{2–10}$$

Let $\psi(t)$ be the *fundamental matrix* of the system in Eq. 2–10, that is, the matrix whose ith column is the solution to Eq. 2–10 with initial condition $\mathbf{z}(0)$ given by the vector which equals one in its ith component but zero elsewhere, that is, $\psi(0) = \mathbf{I}$; then for any vector $\mathbf{z}_0$, $\psi(t)\mathbf{z}_0$ satisfies Eq. 2–10 with $\psi(0)\mathbf{z}_0 = \mathbf{z}_0$. Hence

$$\psi(t)\mathbf{p}'(0) = \mathbf{p}'(t) = \mathbf{p}'(t + T_0) = \psi(t + T_0)\mathbf{p}'(0) = \psi(T_0)\psi(t)\mathbf{p}'(0),$$

which implies that one is an eigenvalue of $\psi(T_0)$; suppose that it is an eigenvalue of multiplicity one. Then for small $|\epsilon|$, it is known [24, p. 352] that Eq. 2–9 has a unique periodic solution $\mathbf{p}(t;\ \epsilon)$ of period T_ϵ, continuous in $(t,\ \epsilon)$. This fact motivates us to attempt to use a perturbation expansion to compute periodic solutions of general nonlinear systems.

EXAMPLE Recall the example of Section 2.6,

$$x'' + x + \epsilon x^3 = 0, \qquad x(0) = 1, \qquad x'(0) = 0$$

for small ϵ; there we computed

$$x(t) = \cos t + \epsilon\left[\frac{\cos 3t}{32} - \frac{\cos t}{32} - \frac{3t \sin t}{8}\right] + O(\epsilon^2).$$

This expansion has a nonperiodic *secular term* in it, namely $t \sin t$, despite the fact that we expect a periodic solution. The result above, however, states that the period is also perturbed by terms in ϵ; thus we try the substitution

$$t = s(1 + a_1\epsilon + a_2\epsilon^2 + \ldots), \qquad x(t) = u_0(s) + \epsilon u_1(s) + \epsilon^2 u_2(s) + \ldots.$$

Substituting into the equation for $x(t)$ and equating powers of ϵ yields

$$\begin{aligned} &u_0'' + u_0 = 0, \qquad u_0(0) = 1, \qquad u_0'(0) = 0 \\ &u_1'' + u_1 = -(2a_1u_0 + u_0^3), \qquad u_1(0) = u_1'(0) = 0, \end{aligned} \tag{2–11}$$

where now the primes indicate differentiation with respect to s. We find

$$u_0(s) = \cos s,$$

so that $u_1(s)$ solves

$$u_1'' + u_1 = -(2a_1\cos s + \cos^3 s),$$

where we will pick a_1 so that no secular term can arise. Since $\cos^3 s = (\cos 3s + 3\cos s)/4$, the function causing the secular term is $-(2a_1\cos s + (3\cos s)/4)$; this disappears if we take $a_1 = -(3/8)$. The equation for $u_1(s)$ becomes

$$u_1'' + u_1 = -\frac{\cos 3s}{4}$$

with solution

$$u_1(s) = \frac{\cos 3s}{32} - \frac{\cos s}{32}$$

satisfying $u_1(0) = u_1'(0) = 0$. This finally yields

$$x(t) = \cos\frac{t}{1 - \frac{3}{8}\epsilon} + \frac{\epsilon}{32}\left[\cos\frac{3t}{1 - \frac{3}{8}\epsilon} - \cos\frac{t}{1 - \frac{3}{8}\epsilon}\right] + O(\epsilon^2)$$

and we see that the period has become

$$T_\epsilon = 2\pi(1 - \tfrac{3}{8}\epsilon) + O(\epsilon^2). \quad \bullet$$

EXERCISE

Fill in the details in the example above.

Additional references: 6, 7, 18, 24.

3

BOUNDARY VALUE PROBLEMS

3.1 INTRODUCTION

Given the differential equation

$$\mathbf{y}' = \mathbf{f}(\mathbf{y}), \qquad \mathbf{y} \text{ in } E^n, \tag{3–1}$$

a boundary value problem is obtained by imposing conditions on the solution at more than one instant in time; most commonly, conditions are imposed at precisely two instants: the only case we shall consider here. Given two continuous mappings $\mathbf{b}_1$ and $\mathbf{b}_2$ of E^n into E^n, we consider *boundary conditions* of the form

$$\mathbf{b}_1(\mathbf{y}(t_1)) = 0, \qquad \mathbf{b}_2(\mathbf{y}(t_2)) = 0;$$

thus the boundary conditions may be nonlinear. For the initial value problem we required n independent conditions, namely the values of the n components of $\mathbf{y}$ at a fixed instant, in order to determine a solution; we expect n conditions to be required in some sense also for the boundary value problem. However, even for very simple equations (e.g., linear), it is possible for a boundary value problem to have no solution, one solution, or infinitely many solutions! Thus boundary value problems are much more complex than initial value problems and the theory, correspondingly, is less developed.

3.2 LINEAR PROBLEMS

We shall call a boundary value problem *linear* when $\mathbf{f}$, $\mathbf{b}_1$, and $\mathbf{b}_2$ are linear. In this case the equation has the form

$$\mathbf{y}' = \mathbf{A}\mathbf{y} \tag{3–2}$$

with conditions

$$\mathbf{M}_1\mathbf{y}(t_1) = \mathbf{k}_1, \qquad \mathbf{M}_2\mathbf{y}(t_2) = \mathbf{k}_2,$$

where $\mathbf{M}_1$ and $\mathbf{M}_2$ are n by n matrices and $\mathbf{k}_1$ and $\mathbf{k}_2$ are vectors.

EXAMPLE Consider the equation

$$x'' + x = 0, \qquad x(0) = 0, \qquad x(T) = x_T,$$

or equivalently,

$$y_1' = y_2$$

$$y_2' = -y_1$$

with conditions $y_1(0) = 0$, $y_1(T) = x_T$, where we have set $y_1 = x$, $y_2 = x'$. This can be written in the general form for linear problems with

$$\mathbf{A} = \begin{pmatrix} 0 & 1 \\ -1 & 0 \end{pmatrix}, \qquad \mathbf{M}_1 = \mathbf{M}_2 = \begin{pmatrix} 1 & 0 \\ 0 & 0 \end{pmatrix}, \qquad \mathbf{k}_1 = \begin{pmatrix} 0 \\ 0 \end{pmatrix}, \qquad \mathbf{k}_2 = \begin{pmatrix} x_T \\ 0 \end{pmatrix}.$$

The general solution to $\mathbf{y}' = \mathbf{A}\mathbf{y}$ is $\mathbf{y} = e^{t\mathbf{A}}\mathbf{y}(0)$, in this case

$$x(t) = y_1(t) = a \sin t$$

where a is not determined by the condition at $t = 0$. Thus we require

$$x_T = x(T) = a \sin T,$$

so that

$$a = \frac{x_T}{\sin T}$$

Therefore we have the *principle of the alternative:* (1) if T is not an integer multiple of π, there is a unique solution to the boundary value problem; (2) if T is an integer multiple of π and $x_T \neq 0$, there are no solutions; (3) if T is an integer multiple of π and $x_T = 0$, there are infinitely many solutions, namely $a \sin t$ for all a. ●

EXERCISE

Verify that $e^{t\mathbf{A}}\mathbf{y}(0)$ has the form given above.

We shall proceed to show that the alternative principle holds for any linear problem; we may for convenience take $t_1 = 0$, $t_2 = T$. The general solution to Eq. 3–2 is

$$\mathbf{y}(t) = e^{t\mathbf{A}}\mathbf{c}$$

for an arbitrary vector $\mathbf{c}$ in E^n. Thus

$$\mathbf{y}(0) = \mathbf{c}$$

and

$$\mathbf{y}(T) = e^{T\mathbf{A}}\mathbf{c},$$

so that the boundary conditions become

$$\begin{aligned} \mathbf{M}_1\mathbf{c} &= \mathbf{k}_1 \\ \mathbf{M}_2 e^{T\mathbf{A}}\mathbf{c} &= \mathbf{k}_2. \end{aligned} \tag{3–3}$$

The questions of existence and uniqueness of the solutions to the boundary value problem are reduced to those questions for solutions of linear algebraic equations. If we write the $2n$-by-n matrix

$$\mathbf{M}(T) = \begin{pmatrix} \mathbf{M}_1 \\ \mathbf{M}_2 e^{T\mathbf{A}} \end{pmatrix},$$

and the vector

$$\mathbf{k} = \begin{pmatrix} \mathbf{k}_1 \\ \mathbf{k}_2 \end{pmatrix},$$

Eq. 3–3 is written

$$\mathbf{M}(T)\mathbf{c} = \mathbf{k}.$$

There exist solutions, therefore, if and only if rank $\mathbf{M}(T)$ = rank $(\mathbf{M}(T), \mathbf{k})$ where $(\mathbf{M}(T), \mathbf{k})$ is the $2n$-by-$(n + 1)$ matrix consisting of $\mathbf{k}$ and the columns of $\mathbf{M}(T)$; the solution is unique if and only if the ranks equal n, while otherwise there are infinitely many solutions or none.

EXERCISE

Analyze the existence and uniqueness of solutions to the linear boundary value problem

$$\begin{aligned} y_1' &= 5y_1 + 2y_2 \\ y_2' &= -2y_1 \end{aligned}$$

with $y_1(0) = 0$, $y_1(T) = x_T$.

Additional references: 43, 61, 64.

3.3 NONLINEAR PROBLEMS; REPRESENTATION

If our boundary value problem is nonlinear, the theory is far more complicated than that above. For some equations, the principle of the alternative derived above has its analogue; we remark that the problem

$$y'' = e^{-y}, \qquad y(0) = y(T) = 0$$

has precisely two solutions if $0 < T < T_0$ where T_0 is a unique critical value, a unique solution if $T = T_0$, and no solutions if $T > T_0$.

EXERCISE

Write the equation $y'' = e^{-y}$, $y(0) = y(T) = 0$ with $T > 0$ as a system in the form $y_1' = y_2$, $y_2' = e^{-y_1}$, $y_1(0) = y_1(T) = 0$. Sketch the vector field for this system. Show that $e^{-y_1} + y_2^2/2 = 1 + y_2(0)^2/2$ along solutions. Deduce that solutions to the boundary value problem start with $y_2(0) < 0$ and that y_2 increases monotonically until $y_2(T) = -y_2(0) > 0$.

EXERCISE

Putting $a \equiv \sqrt{2 + y_2(0)^2}$, and using $y_2' = e^{-y_1} = (a^2/2) - (y_2^2/2)$, deduce an equation for t as a function of y_2; solve this equation to find

$$t = \frac{1}{a} \ln \left\{ \left(\frac{a + y_2(t)}{a - y_2(t)} \right) \left(\frac{a - y_2(0)}{a + y_2(0)} \right) \right\}.$$

Deduce then that

$$y_2(0) = \frac{a\left[1 - \exp\left(\dfrac{aT}{2}\right)\right]}{1 + \exp\left(\dfrac{aT}{2}\right)} \qquad \text{where } a \equiv \sqrt{2 + y_2(0)^2}.$$

EXERCISE

To count the number of solutions to the differential equation, count the number of values of $y_2(0)$ satisfying the relationship immediately above. To do this graphically analyze the two functions $\varphi_1(a) \equiv \sqrt{1 - (2/a^2)}$ and

$$\varphi_2(a) \equiv \frac{1 - \exp\left(\dfrac{-aT}{2}\right)}{1 + \exp\left(\dfrac{-aT}{2}\right)}$$

and the values where $\varphi_1(a) = \varphi_2(a)$. Conclude that there is a value T_0 such that for $T < T_0$ there are two solutions, for $T = T_0$ one solution, and for $T > T_0$ no solutions.

Like linear problems, the nonlinear problem could be reduced formally to a system of algebraic equations via Lie series. Recall from Section 2.4 that the solution to Eq. 3–1 with initial conditions $\mathbf{y}(0) = \mathbf{c}$ can be formally written

$$\mathbf{y} = [e^{t\mathbf{F}}\mathbf{y}]_{\mathbf{y}=\mathbf{c}} \text{ where } \mathbf{F} \equiv \sum_{i=1}^{n} f_i(\mathbf{y})\frac{\partial}{\partial y_i}.$$

Thus we seek a vector $\mathbf{c}$ solving the algebraic system

$$\mathbf{b}_1(\mathbf{c}) = \mathbf{0}, \qquad \mathbf{b}_2([e^{T\mathbf{F}}\mathbf{y}]_{\mathbf{y}=\mathbf{c}}) = \mathbf{0},$$

just as in the linear case. But because for theoretical questions of existence and uniqueness this system is too difficult to handle conveniently, we shall proceed with a somewhat different approach.

Additional references: 53, 65.

3.4 EXISTENCE AND UNIQUENESS

Although in most of the sequel we shall consider a much less general class of problems which arise often in practice, we wish to state a general existence theorem for completeness. Consider the system in Eq. 3–1 with *linear* boundary conditions at $t_1 = 0$, $t_2 = T$; we require that there be n linearly independent boundary conditions. This implies that we can write the conditions as

$$\mathbf{M}\mathbf{y}(0) + \mathbf{N}\mathbf{y}(T) = \mathbf{k},$$

where $\mathbf{M}$ and $\mathbf{N}$ are n by n and $\mathbf{k}$ is in E^n, with $\mathbf{M} + \mathbf{N}$ nonsingular. Let $\partial f_i/\partial y_j$ be continuous and let the matrix $\mathbf{J}$ of those partial derivatives satisfy $\|\mathbf{J}\|_\infty \leq L$ for all $\mathbf{y}$. For arbitrary vector $\mathbf{z}$ in E^n we can solve Eq. 3–1 with the *initial* condition $\mathbf{y}(0) = \mathbf{z}$; this gives a solution $\mathbf{y}(t; \mathbf{z})$. Under these assumptions, if the vector $\mathbf{z}$ solves $\mathbf{M}\mathbf{z} + \mathbf{N}\mathbf{y}(T; \mathbf{z}) = \mathbf{k}$, then $\mathbf{y}(t; \mathbf{z})$ solves the boundary value problem; moreover, if $LT \leq \ln(1+(a/b))$ where $0<a<1$ and $\|(\mathbf{M} + \mathbf{N})^{-1}\mathbf{N}\|_\infty \leq b$, then there is precisely one such solution.

We shall now specialize to a simpler class of systems, namely those equivalent to

$$y'' = f(t, y, y'), \qquad y \in E^1, \qquad t_1 \leq t \leq t_2, \tag{3–4}$$

with general linear boundary conditions, that is,

$$\begin{aligned} ay(t_1) - by'(t_1) = \alpha, &\qquad cy(t_2) + dy'(t_2) = \beta, \\ |a| + |b| \neq 0, &\qquad |c| + |d| \neq 0. \end{aligned} \tag{3–5}$$

Equations of this type arise very often in practice.

Let the function $f(t, y, z)$ be continuous for t in $[t_1, t_2]$ and y, z finite and satisfy a Lipschitz condition in y, z there, that is

$$|f(t, y, z) - f(t, u, w)| \leq L\{|y-u|^2 + |z-w|^2\}^{1/2}, \qquad L < \infty.$$

Suppose $\partial f/\partial y$ and $\partial f/\partial z$ are continuous and satisfy $\partial f/\partial y > 0$, $|\partial f/\partial z| \leq K$ for $t_1 \leq t \leq t_2$, and suppose the coefficients in Eq. 3–5 satisfy $ab \geq 0$, $cd \geq 0$, $|a| + |c| \neq 0$. Then [64, p. 9] the problem given by Eq. 3–4, Eq. 3–5 has a unique solution. This last result includes, for example, those common boundary conditions of the form $y(t_1) = \alpha$, $y(t_2) = \beta$ and those equations independent of y' with $\partial f/\partial y > 0$. We remark that for many problems one does not have a Lipschitz constant independent of y and z; for example, let $f(t, y, z) = e^y$. In this case, one can use the usual trick of letting $g(t, y, z) = e^y$ if $y \leq C$, $g(t, y, z) = \hat{g}(t, y, z) \equiv g_0 y^3 + g_1 y^2 + g_2 y + g_3$ for $C \leq y \leq C+1$ with g_0, g_1, g_2, g_3 chosen so that $\partial g/\partial y$ is continuous and g and $\partial g/\partial y$ vanish at $y = C+1$, with $g(t, y, z) = \hat{g}(t, C+1, 0)$ for $y \geq C+1$. Now g satisfies a Lipschitz condition independent of y, z. Often one can show that, for large C, the solution to $y'' = g(t, y, y')$ satisfies $y \leq C$ and hence $y'' = f(t, y, y')$. This technique works in the following problem.

EXERCISE

Consider $y'' = e^y$, $y(0) = \alpha$, $y(1) = \beta$. Show that there is a unique solution for every value of α, β.

Additional references: 25, 27, 64.

3.5 INTEGRAL EQUATION FORMULATION

Many boundary value problems of a general nature can be stated as integral equations; just as we stated for initial value problems, in Section 2.3, this can be of value for both theoretical and computational purposes. For example, consider the problem

$$\mathbf{y}'' = \mathbf{f}(t, \mathbf{y}), \qquad \mathbf{y} \in E^n, \qquad \mathbf{y}(0) = \mathbf{y}_0, \qquad \mathbf{y}(T) = \mathbf{y}_T. \tag{3–6}$$

Integrating once, we obtain

$$\mathbf{y}'(t) = \mathbf{y}'(0) + \int_0^t \mathbf{f}(\tau, \mathbf{y}(\tau))d\tau.$$

Integrating again yields

$$\mathbf{y}(t) = \mathbf{y}_0 + t\mathbf{y}'(0) + \int_0^t\int_0^s \mathbf{f}(\tau, \mathbf{y}(\tau))d\tau\, ds,$$

and the condition at T yields

$$\mathbf{y}_T = \mathbf{y}_0 + T\mathbf{y}'(0) + \int_0^T\int_0^s \mathbf{f}(\tau, \mathbf{y}(\tau))d\tau\, ds.$$

By interchanging the order of integration, we find

$$\int_0^t\int_0^s \mathbf{f}(\tau, \mathbf{y}(\tau))d\tau\, ds = \int_0^t\int_\tau^t \mathbf{f}(\tau, \mathbf{y}(\tau))ds\, d\tau$$

$$= \int_0^t (t-\tau)\mathbf{f}(\tau, \mathbf{y}(\tau))d\tau,$$

yielding

$$\mathbf{y}'(0) = (\mathbf{y}_T - \mathbf{y}_0)\frac{1}{T} - \frac{1}{T}\int_0^T (T-\tau)\mathbf{f}(\tau, \mathbf{y}(\tau))d\tau$$

and thereby

$$\mathbf{y}(t) = \mathbf{y}_0 + (\mathbf{y}_T - \mathbf{y}_0)\frac{t}{T} + \int_0^T K(t, \tau)\mathbf{f}(\tau, \mathbf{y}(\tau))d\tau, \qquad (3\text{–}7)$$

where

$$K(t, \tau) = \begin{cases} \left(\dfrac{t}{T} - 1\right)\tau & \text{if } 0 \le \tau \le t \\ \left(\dfrac{\tau}{T} - 1\right)t & \text{if } t \le \tau \le T. \end{cases}$$

Thus the boundary value problem in Eq. 3–6 has been converted to the integral equation in Eq. 3–7. One can, under certain assumptions, prove the existence and uniqueness of a solution and, in fact, construct it by applying the Picard iteration method of Section 2.3 to Eq. 3–7. A similar conversion process can be used on other boundary value problems, leading to equations like Eq. 3–7 but with a different definition of the *Green's function* $K(t, \tau)$.

EXERCISE

Use the above techniques to write the solution y of $y'' + ay = f(t, y)$, $y(0) = y_0$, $y(T) = y_T$ in an integral equation involving $f(t, y)$ and an appropriate Green's function under the integral sign.

Additional references: 29, 64.

3.6 VARIATIONAL FORMULATION

Many boundary value problems of the type given in Eq. 3–4 arise in the *calculus of variations;* this allows us to state the problem as an equivalent variational problem.

Consider the problem of minimizing

$$J(\mathbf{y}) = \int_0^1 F(t, \mathbf{y}(t), \mathbf{y}'(t))dt \tag{3–8}$$

over a set of functions with $\mathbf{y}'$ continuous and satisfying $\mathbf{y}(0) = \mathbf{y}_0$, $\mathbf{y}(1) = \mathbf{y}_1$. Then for any function $\mathbf{z}(t)$ with $\mathbf{z}'$ continuous and $\mathbf{z}(0) = \mathbf{z}(1) = \mathbf{0}$, and for any real number α, we must have

$$J(\mathbf{y}) \leq J(\mathbf{y} + \alpha\mathbf{z}) = \int_0^1 F(t, \mathbf{y} + \alpha\mathbf{z}, \mathbf{y}' + \alpha\mathbf{z}')dt.$$

If F is differentiable, then

$$J(\mathbf{y} + \alpha\mathbf{z}) = \int_0^1 \left\{F(t, \mathbf{y}, \mathbf{y}') + \frac{\partial F}{\partial \mathbf{y}}(\alpha\mathbf{z}) + \frac{\partial F}{\partial \mathbf{y}'}(\alpha\mathbf{z}') + o(\alpha)\right\}dt,$$

so that

$$\left[\frac{d}{d\alpha}J(\mathbf{y} + \alpha\mathbf{z})\right]_{\alpha=0} = \int_0^1 \left[\frac{\partial F}{\partial \mathbf{y}} \cdot \mathbf{z} + \frac{\partial F}{\partial \mathbf{y}'}\mathbf{z}'\right]dt = \int_0^1 \left[\frac{\partial F}{\partial \mathbf{y}} - \left(\frac{\partial F}{\partial \mathbf{y}'}\right)'\right]\mathbf{z}\, dt$$

after we have integrated by parts. $J(\mathbf{y}) = J(\mathbf{y} + 0\mathbf{z}) \leq J(\mathbf{y} + \alpha\mathbf{z})$ implies $[dJ(\mathbf{y} + \alpha\mathbf{z})/d\alpha]_{\alpha=0} = 0$. From this, one deduces

$$\frac{\partial F}{\partial \mathbf{y}} - \left(\frac{\partial F}{\partial \mathbf{y}'}\right)' = 0, \tag{3–9}$$

the *Euler equation* for our variational problem. In many cases one can prove [2] that $\mathbf{y}$ solves Eq. 3–9 with $\mathbf{y}(0) = \mathbf{y}_0, \mathbf{y}(1) = \mathbf{y}_1$ if and only if $\mathbf{y}$ minimizes or maximizes $J(\mathbf{y})$; for example [21] if for y in E^1 we have

$$F(t, y, y') = -\frac{y'^2}{2} + G(t, y)$$

with $\partial^2 G/\partial y^2 > 0$, the two problems are equivalent.

EXAMPLE Let $(t, y(t))$ be a curve connecting (0, 0) and (1, 2); that is, let $y(0) = 0$, $y(1) = 2$; we wish to find that curve which when rotated about the t-axis yields the smallest surface area. The formula for surface area is

$$J(y) = 2\pi \int_0^1 y(1 + y'^2)^{1/2}\, dt. \quad \bullet$$

EXERCISE

Derive the Euler equation for the above example.

EXERCISE

Consider the equation

$$y'' = e^y, \qquad y(0) = \alpha, \qquad y(1) = \beta$$

of Section 3.4. Show that this is the Euler equation for the functional

$$J(y) = \int_0^1 \left[-\frac{y'^2}{2} + e^y \right] dt, \qquad y(0) = \alpha, \qquad y(1) = \beta.$$

This approach is both theoretically and computationally useful.

Additional references: 2, 29.

3.7 MONOTONICITY PROPERTIES

Many boundary value problems, especially of the form given by Eq. 3–4 and Eq. 3–5, have monotonicity properties of a very useful type. In general, a mapping **G** is said to be of *monotone kind* if $\mathbf{G}(\mathbf{u}) \geq \mathbf{G}(\mathbf{v})$ implies $\mathbf{u} \geq \mathbf{v}$; this would mean, in particular, that $\mathbf{G}(\mathbf{u}) \geq \mathbf{0}$ would imply $\mathbf{u} \geq \mathbf{u}^*$, where $\mathbf{u}^*$ solves the equation $\mathbf{G}(\mathbf{u}) = \mathbf{0}$. We show that certain boundary value problems are essentially of monotone kind.

Consider differential equations of the form

$$(p(t)y')' = f(t, y, y'), \qquad y \in E^1, \qquad t_1 \leq t \leq t_2 \tag{3–10}$$

with boundary conditions of the type

$$\begin{aligned} ay(t_1) - by'(t_1) &= \alpha \\ cy(t_2) + dy'(t_2) &= \beta, \end{aligned} \tag{3–11}$$

where $p(t)$ is positive and continuously differentiable for $t_1 \leq t \leq t_2$ and where $|a| + |b| \neq 0$, $|c| + |d| \neq 0$. Define related differential and boundary operators **D** and **B** as follows:

$$\begin{aligned} \mathbf{D}(y) &\equiv (-p(t)y')' + f(t, y, y') \\ \mathbf{B}(y) &\equiv \begin{pmatrix} ay(t_1) - by'(t_1) \\ cy(t_2) + dy'(t_2) \end{pmatrix} \in E^2 \end{aligned} \tag{3–12}$$

Our boundary value problem can then be stated as

$$\mathbf{D}(y) = 0, \qquad \mathbf{B}(y) = \begin{pmatrix} \alpha \\ \beta \end{pmatrix};$$

under certain conditions the pair $(\mathbf{D}, \mathbf{B})$ defines a mapping of monotone kind.

Suppose the boundary conditions satisfy $ab \geq 0$ and $cd \geq 0$ and that $\partial f/\partial y$ and $\partial f/\partial y'$ exist and are continuous for $t_1 \leq t \leq t_2$. Let $u(t)$, $v(t)$ be twice continuously differentiable functions of t for $t_1 \leq t \leq t_2$. If $(\partial f/\partial y) > 0$ for $t_1 \leq t \leq t_2$ for all y and y', then $\mathbf{D}(u) \geq \mathbf{D}(v)$ for $t_1 \leq t \leq t_2$ and $\mathbf{B}(u) \geq \mathbf{B}(v)$ imply $u(t) \geq v(t)$ for $t_1 \leq t \leq t_2$. If $f(t, y, y') = f(t, y)$ is independent of y', if $\partial f/\partial y \geq 0$ for $t_1 \leq t \leq t_2$ for all y, and if $|a| + |c| \neq 0$, then for this case also $\mathbf{D}(u) \geq \mathbf{D}(v)$ for $t_1 \leq t \leq t_2$ and $\mathbf{B}(u) \geq \mathbf{B}(v)$ imply $u \geq v$ for $t_1 \leq t \leq t_2$.

EXAMPLE Consider the differential equation

$$y'' = e^y, \qquad 0 \leq t \leq 1,$$

with boundary conditions

$$y(0) = y(1) = 0.$$

Here $f(t, y) = e^y$ is independent of y' and satisfies

$$\frac{\partial f}{\partial y} = e^y > 0.$$

The boundary conditions are such that the differential operator

$$\mathbf{D}(y) = -y'' + e^y$$

along with the boundary operator

$$\mathbf{B}(y) = y$$

give an operator pair of monotone kind. Let $u(t) \equiv 0$. Since $u(0) \geq 0$, $u(1) \geq 0$, and $-u'' + e^u = 1 \geq 0$, it follows that the solution $y(t)$ satisfies

$$y(t) \leq 0. \quad \bullet$$

EXERCISE

Repeat the analysis above for $y'' = y + y^3 + 1$, $y(0) = y(1) = 0$.

Additional references: 25, 26, 27.

PART 2

METHODS OF APPROXIMATE SOLUTION

In spite of the results of Part I concerning existence, uniqueness, and behavior of solutions of differential equations, it is not generally possible to find explicit solutions for problems that arise in practice; in fact, practical problems often refuse to meet the assumptions of theory and thereby remain mute as to the very existence of solutions. We must, then, if we insist on having a solution, compute an approximate one—the word "compute" being used somewhat loosely, to include man's manipulation of the problem before using a digital computer.

To give an idea of the variety of approximation methods available, in the next three chapters we present a sort of brief handbook of several kinds of methods in use today or being proposed for use. In this presentation we attempt to emphasize not the details of each method but the philosophy and important general characteristics of methods of that type. Chapter 4 discusses the standard, basic discrete-variable methods for initial and boundary value problems, including stability problems for initial value problems; integral equation methods for boundary value problems also are covered. Less standard methods for initial value problems are surveyed in Chapter 5, including special multistep, Lie-series, and error-bounding methods. Special methods for boundary value problems, including variational and monotonicity techniques, are presented in Chapter 6. Each of these last two chapters concludes with some personal value judgments on choosing a method for a given problem.

4

BASIC NUMERICAL METHODS FOR INITIAL AND BOUNDARY VALUE PROBLEMS

4.1 INTRODUCTION

Perhaps the basic fact of life when we are making numerical calculations on a computer is that we generally must be content with inexact arithmetic; moreover, computations are performed at such an incomprehensible rate that in a few instants arithmetic and other errors can accumulate and propagate until they completely obscure the desired answer. From a practical point of view, therefore, one of the most important functions of the theoretical analysis of numerical methods for differential equations is to analyze and perhaps to bound the effect that various kinds of errors have on the solution. For this reason we will, in what follows, often emphasize the concept of stability.

Let us consider the problem of computing a numerical solution to the system

$$\mathbf{y}' = \mathbf{f}(t, \mathbf{y}), \qquad \mathbf{y} \in E^n, \qquad a \leq t \leq b \tag{4–1}$$

with initial condition $\mathbf{y}(a) = \mathbf{y}_a$. As we remarked in Section 1.3, this can be converted for conceptual purposes to an autonomous system in which t is considered as another unknown; for computational purposes it is unnecessary and inefficient to solve for values of t, and therefore we shall keep the time-dependent form of our system. A fundamental class of numerical methods is that based on *discrete variables*. We consider a discrete set of instants in time,

$$t_0 = a < t_1 < t_2 < \ldots < t_m = b,$$

and compute a set of vectors $\mathbf{Y}_0, \mathbf{Y}_1, \ldots, \mathbf{Y}_m$ as an approximation to the solution $\mathbf{y}(t)$ evaluated at $t = t_i$; thus we hope

$$\mathbf{Y}_i \sim \mathbf{y}_i \equiv \mathbf{y}(t_i), \qquad i = 0, 1, \ldots, m,$$

where $\sim$ denotes approximate equality. Throughout this chapter we write $h_i \equiv t_{i+1} - t_i$, $h = \max_{0 \le i \le m-1} h_i$. The simplest methods of this type are those that compute the value $\mathbf{Y}_{i+1}$ using the value $\mathbf{Y}_i$ but no other preceding values.

4.2 ONE-STEP METHODS FOR INITIAL VALUE PROBLEMS

Suppose that we let $\mathbf{Y}_0 = \mathbf{y}_a$, and thereafter determine $\mathbf{Y}_{i+1}$ for $i = 0, 1, \ldots, m - 1$ via

$$\mathbf{Y}_{i+1} = \mathbf{Y}_i + h_i \phi(t_i, \mathbf{Y}_i, h_i), \qquad i = 0, 1, \ldots, m - 1, \tag{4–2}$$

where ϕ is some function which depends on $\mathbf{f}$ and is continuous for $a \le t \le b$, $h \le \bar{h}$, and for all $\mathbf{Y}$ where $\bar{h}$ is some positive number. Suppose that there exists a constant M such that

$$\|\phi(t, \mathbf{Y}, k) - \phi(t, \mathbf{Z}, k)\| \le M\|\mathbf{Y} - \mathbf{Z}\|$$

if $a \le t \le b$, $0 \le k \le \bar{h}$, for all $\mathbf{Y}$, $\mathbf{Z}$. The method in Eq. 4–2 is said to be of *order p* if there is a constant L and integer $p > 0$ such that

$$\left\| \phi(t', \mathbf{Y}, k) - \frac{\mathbf{z}(t' + k) - \mathbf{Y}}{k} \right\| \le Lk^p$$

for $a \le t' \le b$, $0 \le k \le \bar{h}$, and all $\mathbf{Y}$ where $\mathbf{z}(t)$ is a sufficiently smooth solution of Eq. 4–1 in $[t', t' + k]$ satisfying $\mathbf{z}(t') = \mathbf{Y}$.

EXERCISE

Show that the one-step method $\mathbf{Y}_{i+1} = Y_i + h\mathbf{f}(t_i, \mathbf{Y}_i)$ is of order one.

If we attempt to compute numerically with Eq. 4–2, we of course make rounding errors in our arithmetic and generate not values $\mathbf{Y}_i$ but values $\mathbf{Z}_i$ satisfying

$$\mathbf{Z}_{i+1} = \mathbf{Z}_i + h_i \phi(t_i, \mathbf{Z}_i, h_i) + \delta_i, \; i = 0, \ldots, m - 1, \tag{4–3}$$

where $\delta = \max_i \|\delta_i\|$ measures the size of our computing errors. If the method in Eq. 4–2 is of order p, then [4, p. 80] the solution $\mathbf{y}(t)$ and the approximation given by Eq. 4–3 satisfy

$$\|\mathbf{y}(t_i) - \mathbf{Z}_i\| \leq \|\mathbf{y}_a - \mathbf{Z}_0\| e^{M(t_i - a)} + \left(Lh^p + \frac{\delta}{h}\right) \frac{e^{M|t_i - a|} - 1}{M}, \qquad (4\text{–}4)$$

if y is smooth enough. Therefore, if $\mathbf{Z}_0 = \mathbf{y}_a$, on a fixed bounded interval of time we get increased accuracy uniformly by decreasing h until the term δ/h begins to dominate Lh^p; for unbounded time intervals, however, the errors may grow exponentially.

EXERCISE

Consider the error in solving $y' = y$, $y(0) = 1$, $y \in E^1$ by $Y_{i+1} = Y_i + hY_i$, $Y_0 = 0$ without rounding error. Find an explicit formula for $E_i \equiv y(ih) - Y_i$. Show that E_i is of order h uniformly for $ih \leq t_0$ fixed, but that the error grows exponentially with ih.

Additional references: 4, 55, 56.

4.3 A ONE-STEP METHOD: TAYLOR SERIES

Suppose that the function $\mathbf{f}(t, \mathbf{y})$ is sufficiently differentiable so that higher derivatives of $\mathbf{y}$ can be computed via Eq. 4–1. That is,

$$\mathbf{y}' = \mathbf{f}(t, \mathbf{y})$$

implies

$$\mathbf{y}'' = \frac{\partial \mathbf{f}}{\partial t} + \frac{\partial \mathbf{f}}{\partial \mathbf{y}} \mathbf{y}' = \frac{\partial \mathbf{f}}{\partial t} + \frac{\partial \mathbf{f}}{\partial \mathbf{y}} \mathbf{f} \equiv \mathbf{f}^{(1)}(t, \mathbf{y}), \qquad \mathbf{y}''' = \frac{d}{dt} \mathbf{f}^{(1)}(t, \mathbf{y}) \equiv \mathbf{f}^{(2)}(t, \mathbf{y}),$$

and so on until

$$\mathbf{y}^{(p)} = \mathbf{f}^{(p-1)}(t, \mathbf{y}).$$

By Taylor's theorem, with $\mathbf{y}_i^{(j)} = \mathbf{y}^{(j)}(t_i)$,

$$\mathbf{y}_{i+1} = \mathbf{y}_i + h_i \mathbf{y}_i^1 + \frac{h_i^2}{2!} \mathbf{y}_i^2 + \cdots + \frac{h_i^p}{p!} \mathbf{y}_i^{(p)} + O(h_i^{p+1})$$

$$= \mathbf{y}_i + h_i \left[\mathbf{f}(t, \mathbf{y}_i) + \frac{h_i}{2} \mathbf{f}^{(1)}(t, \mathbf{y}_i) + \cdots + \frac{h_i^{p-1}}{p!} \mathbf{f}^{(p-1)}(t, \mathbf{y}_i) \right] + O(h_i^{p+1}).$$

Define

$$\phi(t, \mathbf{y}, h) \equiv \mathbf{f}(t, \mathbf{y}) + \frac{h}{2}\mathbf{f}^{(1)}(t, \mathbf{y}) + \ldots + \frac{h^{p-1}}{p!}\mathbf{f}^{(p-1)}(t, \mathbf{y}).$$

EXERCISE

Show that this defines a method of order p and hence the preceding theory applies.

If $p = 1$, we have the well-known method of Euler,

$$\mathbf{Y}_{i+1} = \mathbf{Y}_i + h_i \mathbf{f}(t_i, \mathbf{Y}_i).$$

EXAMPLE Consider

$$y' = y^2, \qquad y \in E^1.$$

Then

$$y'' = 2yy' = 2y^3 \equiv f^{(1)}(t, y), \qquad y''' = 6y^2y' = 6y^4 \equiv f^{(2)}(t, y),$$

and in general

$$y^{(j)} = j!\,y^{j+1} \equiv f^{(j-1)}(t, y).$$

A method of order three for this example results from

$$Y_{i+1} = Y_i + h_i[Y_i^2 + h_i Y_i^3 + h_i^2 Y_i^4]. \quad \bullet$$

Additional references: 4, 27, 44, 55, 56, 81.

4.4 A ONE-STEP METHOD: RUNGE-KUTTA

A possible disadvantage of Taylor-series methods is that they require the functions $\mathbf{f}^{(j)}(t, \mathbf{y})$ for some $j \geq 1$; that is, they require exact formal differentiation. There are now computer programs which can accomplish this: however, if $\mathbf{f}(t, \mathbf{y})$ is very complicated or is given empirically, perhaps by table, formal differentiation may be inefficient or even impossible. Methods of the Runge-Kutta type attempt to produce the equivalent of a Taylor-series expansion by approximately evaluating $\mathbf{f}(t, \mathbf{y})$ at values of t between t_i and

t_{i+1} instead of computing higher derivatives formally. In general, given the value $\mathbf{Y}_i$, one sets

$$\mathbf{k}_j = \mathbf{f}(t_i + \alpha_j h_i,\ \mathbf{Y}_i + h_i \sum_{l=1}^{j-1} \beta_{jl}\mathbf{k}_l), \qquad j = 1, 2, \ldots, J,$$

$$\mathbf{Y}_{i+1} = \mathbf{Y}_i + h_i \sum_{j=1}^{J} \gamma_j \mathbf{k}_j,$$

where the constants α_j, β_{jl}, γ_j are chosen so that the series expansion of $\mathbf{Y}_i + h_i\sum_{j=1}^{J}\gamma_j\mathbf{k}_j$ in powers of h_i matches the Taylor-series expansion for $\mathbf{Y}_{i+1}$ in h_i to as high a degree as possible with a low amount of computational complexity.

EXAMPLE Perhaps the most widely used Runge-Kutta method is of order four; the usual form is as follows:

$$\begin{aligned}
\mathbf{k}_1 &= \mathbf{f}(t_i, \mathbf{Y}_i),\\
\mathbf{k}_2 &= \mathbf{f}(t_i + \tfrac{1}{2}h_i,\ \mathbf{Y}_i + \tfrac{1}{2}h_i\mathbf{k}_1),\\
\mathbf{k}_3 &= \mathbf{f}(t_i + \tfrac{1}{2}h_i,\ \mathbf{Y}_i + \tfrac{1}{2}h_i\mathbf{k}_2),\\
\mathbf{k}_4 &= \mathbf{f}(t_i + h_i,\ \mathbf{Y}_i + h_i\mathbf{k}_3)\\
\mathbf{Y}_{i+1} &= \mathbf{Y}_i + \tfrac{1}{6}h_i(\mathbf{k}_1 + 2\mathbf{k}_2 + 2\mathbf{k}_3 + \mathbf{k}_4).
\end{aligned}$$

With the fourth-order formula in this form, it is possible with a little extra work to estimate the error we make locally in accepting the computed value of $\mathbf{Y}_{i+1}$. Let

$$\begin{aligned}
\mathbf{k}_5 &= \mathbf{f}(t_i + \tfrac{3}{4}h_i,\ \mathbf{Y}_i + \tfrac{1}{32}h_i(5\mathbf{k}_1 + 7\mathbf{k}_2 + 13\mathbf{k}_3 - \mathbf{k}_4))\\
\mathbf{d} &= \tfrac{2}{3}h_i(-\mathbf{k}_1 + 3\mathbf{k}_2 + 3\mathbf{k}_3 + 3\mathbf{k}_4 - 8\mathbf{k}_5).
\end{aligned}$$

Then $\mathbf{d}$ is an $O(h_i^5)$ approximation to the last matching term (i.e., the term in h_i^4) in the actual and approximated series expansions for $\mathbf{Y}_{i+1}$ and can therefore be used to help control the error by keeping h_i small enough that d does not become "too large." ●

Additional references: 4, 17, 27, 44, 55, 56, 77, 91, 92, 109.

4.5 LINEAR MULTISTEP METHODS FOR INITIAL VALUE PROBLEMS

One-step methods apparently have a disadvantage in that they make no use of the recent history of the solution prior to t_i; multistep methods are designed to use this history intelligently. We could try to devise a scheme to solve Eq. 4–1 by replacing the derivative $\mathbf{y}'$ by some finite difference approximation $\mathbf{y}_i' \sim (1/h)(\alpha_k \mathbf{y}_{i+1} + \alpha_{k-1}\mathbf{y}_i + \ldots + \alpha_0 \mathbf{y}_{i+1-k})$ and then compute via

$$\alpha_k \mathbf{Y}_{i+1} = -(\alpha_{k-1}\mathbf{Y}_i - \ldots - \alpha_0 \mathbf{Y}_{i+1-k}) + h\mathbf{f}(t_i, \mathbf{Y}_i),$$

where for simplicity we assume $h_i = h$, $i = 0, \ldots, m-1$ throughout this section. We might also note that $\mathbf{y}(t_{i+1}) = \mathbf{y}(t_i) + \int_{t_i}^{t_{i+1}} \mathbf{f}(t, \mathbf{y}(t))dt$ and replace the integral by a quadrature formula, yielding

$$\mathbf{Y}_{i+1} = \mathbf{Y}_i + h[\beta_k \mathbf{f}(t_{i+1}, \mathbf{Y}_{i+1}) + \beta_{k-1}\mathbf{f}(t_i, \mathbf{Y}_i) + \ldots + \beta_0 \mathbf{f}(t_{i+1-k}, \mathbf{Y}_{i+1-k})].$$

Combining these two approaches, we get the general *linear k-step* method

$$\alpha_k \mathbf{Y}_{i+1} + \alpha_{k-1}\mathbf{Y}_i + \ldots + \alpha_0 \mathbf{Y}_{i+1-k} = h[\beta_k \mathbf{f}_{i+1} + \beta_{k-1}\mathbf{f}_i + \ldots + \beta_0 \mathbf{f}_{i+1-k}] \tag{4–5}$$

where $\mathbf{f}_j \equiv \mathbf{f}(t_j, \mathbf{Y}_j)$ and the starting values $\mathbf{Y}_0, \mathbf{Y}_1, \ldots, \mathbf{Y}_{k-1}$ are computed by some other method, generally as in Section 4.2. We assume of course that $\alpha_k \neq 0$.

We remark that the ease of implementing Eq. 4–5 depends on whether or not $\beta_k = 0$. If $\beta_k = 0$, Eq. 4–5 immediately yields the value of $\mathbf{Y}_{i+1}$; in this case the method is called *explicit*. If $\beta_k \neq 0$, however, we must solve a generally nonlinear equation for $\mathbf{Y}_{i+1}$; such methods are called *implicit*.

The method in Eq. 4–5 is said to be of *order p* if, for all sufficiently differentiable functions $\mathbf{z}(t)$, we have

$$\alpha_k \mathbf{z}(t+h) + \alpha_{k-1}\mathbf{z}(t) + \ldots + \alpha_0 \mathbf{z}(t+h-kh) - h[\beta_k \mathbf{z}'(t+h) + \beta_{k-1}\mathbf{z}'(t) + \ldots$$
$$+ \beta_0 \mathbf{z}'(t+h-kh)] = C_{p+1} h^{p+1} \mathbf{z}^{(p+1)}(t) + O(h^{p+2}), \qquad C_{p+1} \neq 0.$$

EXERCISE

Compare the two definitions of order for a one-step method.

Define the polynomial

$$\rho(z) = \alpha_k z^k + \alpha_{k-1} z^{k-1} + \ldots + \alpha_0.$$

We shall call a method *D-stable* (stable in the sense of Dahlquist) if and only if all roots z_i of $\rho(z) = 0$ satisfy $|z_1| \leq 1$ and $|z_1| = 1$ implies $\rho'(z_i) \neq 0$. Suppose that $\mathbf{f}(t, \mathbf{y})$ is continuous for all $\mathbf{y}$ and $a \leq t \leq b$ and that there exists a constant M such that $\|\mathbf{f}(t, \mathbf{y}) - \mathbf{f}(t, \mathbf{z})\| \leq M\|\mathbf{y} - \mathbf{z}\|$ for all $\mathbf{y}$, $\mathbf{z}$

and $a \leq t \leq b$. Suppose the solution $\mathbf{y}(t)$ of Eq. 4–1 has continuous derivatives of order up to and including $p + 1$, where the method in Eq. 4–5 is of order p. Suppose the method is D-stable and that we compute via Eq. 4–5, except that we allow at each step additional errors δ_i. Then [4, p. 56], for smooth enough solutions,

$$\|\mathbf{Y}_i - \mathbf{y}(t_i)\| \leq \frac{G}{1 - hLM}\left[(1 + hLM)E + \frac{|t_i - a|}{|\alpha_k|}\left(\frac{\delta}{h} + Kh^p\right)\right]e^{P(t_i - a)}$$

where G and P are constants depending on the method but independent of h for small h, $L = |(\beta_k/\alpha_k)|$, $\delta \geq \max_i\{\|\delta_i\|\}$, $E \geq \max\{\|\mathbf{Y}_0 - \mathbf{y}(t_0)\|, \|\mathbf{Y}_1 - \mathbf{y}(t_1)\|, \ldots, \|\mathbf{Y}_{k-1} - \mathbf{y}(t_{k-1})\|\}$. Thus, if the starting values are accurate to $O(h^p)$ the numerical solution is accurate uniformly to $O(h^p)$ over a bounded interval, except for the effect of the computational errors described by the δ term.

EXAMPLES

(i) $\mathbf{Y}_{i+1} = \mathbf{Y}_i + h\mathbf{f}_i$, explicit of order one (Euler's method).

(ii) $\mathbf{Y}_{i+1} = \mathbf{Y}_i + \frac{1}{2}h(3\mathbf{f}_i - \mathbf{f}_{i-1})$, explicit of order two.

(iii) $\mathbf{Y}_{i+1} = \mathbf{Y}_{i-1} + 2h(\frac{7}{6}\mathbf{f}_i - \frac{1}{3}\mathbf{f}_{i-1} + \frac{1}{6}\mathbf{f}_{i-2})$, explicit of order three.

(iv) $\mathbf{Y}_{i+1} = \mathbf{Y}_i + \frac{1}{2}h(\mathbf{f}_i + \mathbf{f}_{i+1})$, implicit of order two (trapezoidal rule).

(v) $\mathbf{Y}_{i+1} = \mathbf{Y}_i + \frac{1}{12}h(5\mathbf{f}_{i+1} + 8\mathbf{f}_i - \mathbf{f}_{i-1})$, implicit of order three.

(vi) $\mathbf{Y}_{i+1} = \mathbf{Y}_{i-1} + 2h(\frac{1}{6}\mathbf{f}_{i+1} + \frac{2}{3}\mathbf{f}_i + \frac{1}{6}\mathbf{f}_{i-1})$, implicit of order four.

(vii) $\frac{11}{6}\mathbf{Y}_{i+1} - 3\mathbf{Y}_i + \frac{3}{2}\mathbf{Y}_{i-1} - \frac{1}{3}\mathbf{Y}_{i-2} = h\mathbf{f}_{i+1}$, implicit of order three.

These methods are D-stable. ●

EXERCISE

Show that all seven of the above methods are D-stable.

EXERCISE

Show that the method of (iv) is of order two.

Additional references: 4, 27, 32, 44, 55, 56.

4.6 PREDICTOR-CORRECTOR METHODS FOR INITIAL VALUE PROBLEMS

Since implicit multistep methods require solution of an equation, on first glance it appears they would never be used; but if implicit methods are combined suitably with explicit methods they become of great value. Suppose, given $\mathbf{Y}_i$, we use an explicit formula of order $p - 1$ to compute a predicted value $\hat{\mathbf{Y}}_{i+1}$; use this value of $\hat{\mathbf{Y}}_{i+1}$ on the right-hand side of Eq. 4–5 in an implicit formula to compute a corrected value of $\mathbf{Y}_{i+1}$, where we suppose that the order of the implicit formula is p; accept this as our value of $\mathbf{Y}_{i+1}$; and proceed to $\mathbf{Y}_{i+2}$ and so on. Despite the fact that the order p implicit equation was not solved exactly but was only used as a *corrector* after the $p - 1$ order explicit *predictor*, the order of the overall computation is p, not $p - 1$. Thus one can use implicit formulas without solving them precisely.

EXERCISE

Show that a method of order two results from using the formula of (i) as a predictor and that of (iv) as a corrector applied only once.

EXAMPLE Consider the combination of examples (ii) and (v) of Section 4.5. As a predictor we take the second-order method

$$\hat{\mathbf{Y}}_{i+1} = \mathbf{Y}_i + \tfrac{1}{2}h(3\mathbf{f}(t_i, \mathbf{Y}_i) - \mathbf{f}(t_{i-1}, \mathbf{Y}_{i-1})),$$

while as a corrector we take the third-order method

$$\mathbf{Y}_{i+1} = \mathbf{Y}_i + \tfrac{1}{12}h(5\mathbf{f}(t_{i+1}, \hat{\mathbf{Y}}_{i+1}) + 8\mathbf{f}(t_i, \mathbf{Y}_i) - \mathbf{f}(t_{i-1}, \mathbf{Y}_{i-1})),$$

so that the result is of order three. ●

Additional references: 4, 27, 44, 55, 56.

4.7 STABILITY CONCEPTS FOR INITIAL VALUE PROBLEMS

We have introduced the concept of D-stability and have indicated how it allows us to deduce error bounds and convergence results over bounded time intervals. The value of D-stability in general can be seen by looking at the simple equation

$$y' = \lambda y, \qquad y \in E^1, \qquad y(0) = 1, \qquad \lambda \text{ constant},$$

with solution $y(t) = e^{\lambda t}$. Using a D-stable linear multistep method, the numerical solution solves the linear difference equation

$$(\alpha_k - \lambda h\beta_k)Y_{i+1} + (\alpha_{k-1} - \lambda h\beta_{k-1})Y_i + \ldots + (\alpha_0 - \lambda h\beta_0)Y_{i+1-k} = 0.$$

Therefore

$$Y_i = \sum_{j=0}^{k} a_j r_j(\lambda h)^i$$

where a_j are constants and $r_j(\lambda h)$ are the roots of $(\alpha_k - \lambda h\beta_k)z^k + \ldots + (\alpha_0 - \lambda h\beta_0) = 0$. For h small these roots are near the roots of $\rho(z)$—see Section 4.5—so we have

$$r_j(\lambda h) = r_j(0) + o(1)$$

and, if $i = t/h$, $|r_j(\lambda h)|^i = O(e^{|t|})$. Since the error $Y_i - y(t_i)$ and Y_i satisfy difference equations that are identical except for the added presence of a $O(h^{p+1})$ term on the right, one can deduce that the error is bounded by $O(e^t h^p)$ for small h.

EXERCISE

Repeat the above analysis for Euler's method and the trapezoidal rule.

Consider, however, the case of $\lambda < 0$, for which the true solution tends to zero as t tends to infinity; we would hope that for small h this would be true of the numerical solution also. If we use the D-stable method

$$Y_{i+1} = Y_{i-1} + 2hf_i,$$

we get $Y_{i+1} = Y_{i-1} + 2h\lambda Y_i$ and the root $r_1(\lambda h) = -\sqrt{1 + \lambda^2 h^2} + \lambda h < -1$ for all $h > 0$; our numerical solution will oscillate wildly if i is large! (See Figure 4–1.) The problem is that the root $r_1(0) = -1$ on the unit circle drifts outside for $\lambda < 0$ and any $h > 0$. For some problems, then, we need methods that are more than D-stable. For a k-step linear multistep method with k even, if the maximum order of accuracy $k + 2$ is attained then all of the roots $r_i(0)$ lie precisely *on* the unit circle. It has been shown [13] that one can derive *for each h* a different method, also of order $k + 2$, whose roots are $r_i(0) + O(h)$ and lie *inside* the unit circle and are such that the $r_i(\lambda h)$ for that method also are inside the circle for small h. We shall not consider this approach here but instead examine a stronger stability concept.

t	0	10	15	20	50	100	200	201
e^{-t}	1	$10^{-4.3}$	$10^{-6.4}$	$10^{-8.6}$	10^{-21}	10^{-43}	10^{-86}	$10^{-86.4}$
Numerical solution	1	$10^{-4.3}$	$10^{-4.4}$	$10^{-4.2}$	10^{-3}	10^{-1}	$10^{3.1}$	$-10^{3.1}$

FIGURE 4–1

For $p \geq 1$, a method of order p always has $\rho(1) = 0$. We shall say a method is *S-stable* (strongly stable) if all k roots of $\rho(z)$ other than $z = 1$ lie strictly inside the unit circle. Note that an S-stable method is certainly D-stable, and thus our earlier theory applies. Moreover, for an S-stable method, for h small enough, all roots $r_i(\lambda h)$—except for $r_1(\lambda h)$ where $r_1(0) = 1$—lie inside the unit circle, and hence their effect on the numerical solution decays rapidly. For $\lambda < 0$ and h small, the numerical solution tends to zero.

EXERCISE

Show that the methods (i), (ii), (iv), (v), (vii), of Section 4.5 are S-stable, and that (iii) and (vi) are not.

More generally, if we solve

$$\mathbf{y}' = \mathbf{A}\mathbf{y}, \qquad \mathbf{y} \in E^n,$$

where the eigenvalues of $\mathbf{A}$ have negative real parts, an S-stable method will yield, for small h, a numerical solution tending to zero. How small h must be is determined by the requirement that $r_j(\lambda h) = r_j(0) + O(|\lambda h|)$ lie in the unit circle; if $|\lambda|$ is large for any eigenvalue λ of $\mathbf{A}$, one would expect to need h smaller than if $|\lambda|$ were small for all eigenvalues λ.

EXERCISE

Determine how small h must be for the numerical solution of the following two systems to tend to zero, using method (vii) of Section 4.5.

(i) $$\begin{pmatrix} Y_1 \\ Y_2 \end{pmatrix}' = \begin{pmatrix} -1 & 0 \\ 0 & -1 \end{pmatrix} \begin{pmatrix} Y_1 \\ Y_2 \end{pmatrix}$$

(ii) $$\begin{pmatrix} Y_1 \\ Y_2 \end{pmatrix}' = \begin{pmatrix} -100 & 0 \\ 0 & -1 \end{pmatrix} \begin{pmatrix} Y_1 \\ Y_2 \end{pmatrix}.$$

We mentioned in Section 2.2 that often, especially in physical problems, we must solve equations in which one eigenvalue has a negative real part of large magnitude while another has a negative real part of small magnitude. Since the effect of the eigenvalue with a large negative real part is transient and dies off very quickly, it is the effect of the other that concerns us for large t. For the numerical solution on the other hand, the stability analysis above tells us that for large t it is the physically unimportant large eigenvalue which determines how small h need to be to guarantee stability. The continuing artificial influence of the large eigenvalue is very unpleasant. In a later section we shall discuss the concept of "A-stability" for such problems.

A similar analysis of the meaning and value of stability holds for one-step methods as well; consider the fourth-order Runge-Kutta method described in Section 4.4 applied to the equation

$$\mathbf{y}' = \mathbf{A}\mathbf{y}, \qquad \mathbf{y} \in E^n.$$

We find

$$\mathbf{Y}_{i+1} = (1 + h\mathbf{A} + \tfrac{1}{2}h^2\mathbf{A}^2 + \tfrac{1}{6}h^3\mathbf{A}^3 + \tfrac{1}{24}h^4\mathbf{A}^4)\mathbf{Y}_i \equiv P(h\mathbf{A})\mathbf{Y}_i.$$

If the eigenvalues of $\mathbf{A}$ have negative real parts, then for small enough h the numerical solution will tend to zero, just like the actual solution, since the eigenvalues of $P(h\mathbf{A})$ will lie strictly within the unit circle. Thus we could say that the Runge-Kutta method is S-stable; the remarks in the preceding paragraph concerning what determines how small h must be certainly apply here as well.

Additional references: 4, 27, 30, 31, 32, 34, 44, 55, 56.

4.8 ON CHOOSING THE STEP SIZE FOR INITIAL VALUE PROBLEMS

Given that for some reason one has decided to use a certain formula to obtain a numerical solution to a differential equation in a certain form, how small should h be? Perhaps more importantly, how should we change h from step to step: that is, how should the values of h_i be determined? We of course have no good simple answers for these questions. As we stated in Section 4.4, one can estimate the local error in certain forms of the Runge-Kutta methods and thus vary the step size to keep *local* error within a certain tolerance. (See also Section 9.2.) Variable-step versions of multistep and predictor-corrector methods exist in which we can estimate error also, but again this is an estimate of local error [69, 83, 103]. With the exception of the error-bounding techniques discussed in the next chapter, we know of no methods for estimating well the overall propagated errors. This is very critical, since decreased step sizes magnify the effect of rounding errors while they increase local accuracy; at some point the disadvantages brought by reducing h more than nullify the advantages. An example computed elsewhere [4, p. 71] shows that in solving

$$y' = 1 - y, \qquad y(0) = 2$$

by the formula

$$Y_{i+1} = Y_i + \tfrac{1}{2}h(3f_i - f_{i-1})$$

with Y_0 and Y_1 exact, the curves representing theoretical error $y(t_i) - Y_i$, round-off error, and actual error $y(t_i)$ minus the *computed* Y_i at any value of $t \leq 5$ are as in Figure 4–2. Thus we see the need of keeping h from getting too small in addition to the need of keeping it from getting too large! Much more theory regarding the monitoring of the size of h needs to be developed.

The question of choosing a basic method is much more difficult; we shall leave this to the end of the next chapter.

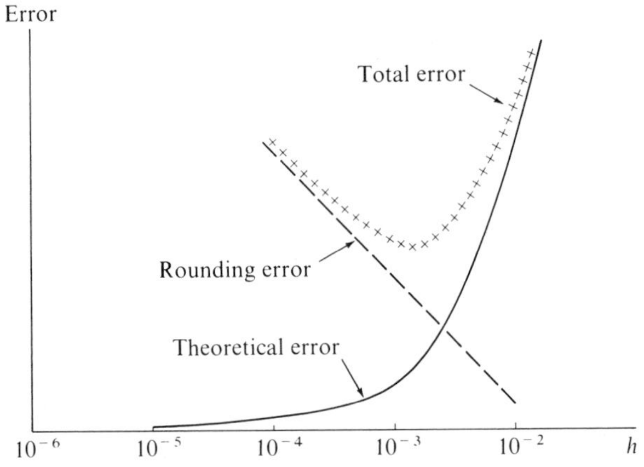

FIGURE 4–2

EXERCISE

Conduct a numerical experiment like the one described above, using Euler's method to solve $y' = y$, $y(0) = 1$.

Additional references: 4, 44, 55, 56, 59, 91, 92, 109.

4.9 FINITE DIFFERENCE METHODS FOR BOUNDARY VALUE PROBLEMS

Consider the numerical solution of a time-dependent form of the two-point boundary value problem discussed in Section 3.4,

$$\mathbf{y}' = \mathbf{f}(t, \mathbf{y}), \qquad \mathbf{y} \in E^n, \qquad 0 \leq t \leq T \tag{4–6}$$

with linear boundary conditions

$$\mathbf{M}\mathbf{y}(0) + \mathbf{N}\mathbf{y}(T) = \mathbf{k}, \tag{4–7}$$

with $\mathbf{M}$ and $\mathbf{N}$ being n-by-n matrices such that $\mathbf{M} + \mathbf{N}$ is nonsingular, and $\mathbf{k}$ being a member of E^n. As is true for initial value problems, we can write down difference formulas involving the approximate solution $\mathbf{Y}_i$ at grid points t_i with $0 = t_0 < t_1 < \ldots < t_m = T$, $t_{i+1} - t_i = h$ for convenience; these formulas can be based on approximate differentiation and integration as discussed in Section 4.5. For our boundary value problem, however, insufficient data is given at $t = 0$ to determine $\mathbf{Y}_0$ directly and then $\mathbf{Y}_1$, et cetera; we produce instead a *system* of equations to solve.

For example, consider the difference approximation

$$\mathbf{Y}_{i+1} = \mathbf{Y}_i + h\mathbf{f}(t_i + \tfrac{1}{2}h, \tfrac{1}{2}(\mathbf{Y}_{i+1} + \mathbf{Y}_i)), \qquad i = 0, \ldots, m-1 \tag{4–8}$$

with boundary conditions

$$\mathbf{M}\mathbf{Y}_0 + \mathbf{N}\mathbf{Y}_m = \mathbf{k}. \tag{4–9}$$

We have $m + 1$ sets of n equations to solve for the $n(m + 1)$ components of the solution at the grid points. Let $\mathbf{f}(t, \mathbf{y})$ and $(\partial\mathbf{f}/\partial\mathbf{y}) \equiv \mathbf{J}(t, \mathbf{y})$ be continuous for all $\mathbf{y}$ and $0 \leq t \leq T$ with $\|\mathbf{J}(t, \mathbf{y})\|_\infty \leq L$; suppose $b = \max\{\|(\mathbf{M}+\mathbf{N})^{-1}\mathbf{M}\|_\infty, \|(\mathbf{M}+\mathbf{N})^{-1}\mathbf{N}\|_\infty\}$, $0 < a < 1$ and $LT \leq \ln(1 + (a/b))$. Then, as we saw in Section 3.4, the boundary value problem given by Eq. 4–6 and Eq. 4–7 has a unique solution. Moreover [64, p. 94–98], the system given by Eq. 4–8 and Eq. 4–9 also has a unique solution for all h, and as h tends to zero the numerical solutions tend to the solution of the boundary value problem; if, in addition, $\mathbf{f}(t, \mathbf{y})$ is twice continuously differentiable with respect to $\mathbf{y}$, the numerical solutions are accurate to $O(h^2)$.

As we stated in Section 3.4, many problems of interest have the form

$$\mathcal{L}(y) \equiv -y'' + f(t, y, y') = 0, \qquad y \in E^1, \qquad 0 \leq t \leq T$$

with $y(0) = \alpha$, $y(T) = \beta$, where $f(t, y, z)$, $\partial f/\partial y$, and $\partial f/\partial z$ are continuous,

$|\partial f/\partial z| \leq L$ and $0 < l \leq \partial f/\partial y \leq u < \infty$ for $0 \leq t \leq T$ and all y, z. The differential equation is approximated by

$$\mathcal{L}_h(Y_i) \equiv -\frac{1}{h^2}(Y_{i+1} - 2Y_i + Y_{i-1}) + f\left(t_i, Y_i, \frac{1}{2h}(Y_{i+1} - Y_{i-1})\right) = 0, \qquad i = 1, \ldots, m-1. \tag{4–10}$$

The preceding results about existence and convergence are essentially valid here with the additional stability result that for $h \leq 2/L$, for all Y_i, Z_i,

$$|Y_i - Z_i| \leq P\{\max(|Y_0 - Z_0|, |Y_m - Z_m|) + \max_{1\leq j\leq m-1}|\mathcal{L}_h(Y_j) - \mathcal{L}_h(Z_j)|\},$$

where $P = \max(1, 1/l)$. Since $\mathcal{L}_h(y(t_i)) = O(h^2)$, this stability result implies that if one has $Y_0 = \alpha$, $Y_m = \beta$, but $\mathcal{L}_h(Y_i) = \delta_i$ instead of $\mathcal{L}_h(Y_i) = 0$, it follows that

$$|Y_i - y(t_i)| = O(h^2 + |\delta_i|).$$

A common special type of equation has the form

$$y'' = f(t, y).$$

Since

$$\frac{y_{i+1} - 2y_i + y_{i-1}}{h^2} = y_i'' + \frac{h^2}{12}y^{(4)} + O(h^4),$$

the formula in Eq. 4–10 can be modified in this case without increasing the complexity to give

$$\begin{aligned}\frac{1}{h^2}[Y_{i+1} - 2Y_i + Y_{i-1}] &= f_i + \tfrac{1}{12}(f_{i+1} - 2f_i + f_{i-1}) \\ &= \tfrac{1}{12}(f_{i+1} + 10f_i + f_{i-1}),\end{aligned}$$

a method yielding an $O(h^4)$ solution. This method is widely used. For example [87], step sizes of $h = \pi/10$ and $\pi/20$ yield solutions with maximum errors of 1.2×10^{-5} and 7.4×10^{-7}, respectively, when applied to the problem

$$y'' = y^3 - (1 + \sin^2 t)\sin t, \qquad y(0) = y(\pi) = 0$$

having exact solution

$$y(t) = \sin t.$$

We shall not discuss here the practical problems of computing a solution to the nonlinear equations in Eq. 4–8, Eq. 4–9, or Eq. 4–10; despite the importance of this problem, we cannot do it justice in a brief presentation. Much effort has been and still is being brought to bear on the computational algorithms designed to solve such problems; this is a vitally important area of research that is fundamental to the efficient numerical solution of boundary value problems by difference methods. New methods undoubtedly will be developed to provide more algorithms in addition to the well-known direct or iterative methods such as Newton, nonlinear Gauss-Siedel, et cetera [84, 86, 105].

EXAMPLE To solve

$$y'' = y, \qquad y(0) = 1, \qquad y(1) = e \equiv \exp(1),$$

having exact solution $y(t) = e^t$, we use the above method. Our equations become

$$-Y_{i+1} + (2+h^2)Y_i - Y_{i-1} = 0, \qquad Y_0 = 1, \qquad Y_m = e, \qquad i = 1, 2, \ldots, m-1.$$

The error satisfies, for $h = .05$,

$$|Y_i - y(t_i)| \leq .0003. \quad \bullet$$

Additional references: 4, 27, 44, 45, 55, 56, 64, 73.

4.10 SHOOTING METHODS FOR BOUNDARY VALUE PROBLEMS

We consider again the solution of the problem defined by Eq. 4–6 and Eq. 4–7 under the assumptions given following Eq. 4–9, which guarantee the existence and uniqueness of a solution, as we saw earlier in Section 3.4. The method used to analyze existence was to consider the solution $\mathbf{y}(t; \mathbf{z})$ to the problem in Eq. 4–6 with *initial* condition $\mathbf{y}(0) = \mathbf{z}$; the problem then was to solve the equation

$$\mathbf{Mz} + \mathbf{Ny}(T; \mathbf{z}) = \mathbf{k} \tag{4–11}$$

for the unique unknown vector $\mathbf{z}$. Clearly this theoretical method lends itself to computational uses easily.

In the *shooting* method, we guess an initial vector $\mathbf{z}$ and then "shoot" towards the other end of the given t-interval by solving the initial value problem; when we reach $t = T$ we modify $\mathbf{z}$, if necessary, to solve Eq. 4–11 more accurately. We thus shoot first and ask questions later.

EXERCISE

Consider solving $y'' + y = 0$, $y(0) = 0$, $y(1) = \sin 1$ by shooting. Find the exact solution $y(t;\, \alpha)$ of $y'' + y = 0$, $y(0) = 0$, $y'(0) = \alpha$, and then solve the equation $y(1; \alpha) = \sin 1$ for α to find a solution to the boundary value problem.

Any numerical method for initial value problems can be used to generate an approximate solution $\mathbf{Y}_i(\mathbf{z})$; if the method is stable and of order p, then, ignoring rounding errors,

$$\mathbf{Y}_T(\mathbf{z}) = \mathbf{y}(T;\, \mathbf{z}) + O(h^p),$$

where by $\mathbf{Y}_T(\mathbf{z})$ we mean the numerical solution to the initial value problem evaluated at $t = T$. Instead of solving Eq. 4–11, we seek instead to solve

$$\mathbf{M}\mathbf{z} + \mathbf{N}\mathbf{Y}_T(\mathbf{z}) = \mathbf{k} \tag{4–12}$$

for the initial value $\mathbf{z}$.

EXERCISE

Solve the problem of the preceding exercise again, only this time find α such that $Y(1;\, \alpha) = \sin 1$ where $Y(ih;\, \alpha)$ is the solution to $y'' + y = 0$, $y(0) = 0$, $y'(0) = \alpha$ obtained via Euler's method.

It can be shown [64, p. 57] that the sequence of vectors $\mathbf{z}^j$ defined by

$$\mathbf{z}^{j+1} = \mathbf{z}^j - (\mathbf{M} + \mathbf{N})^{-1}[\mathbf{M}\mathbf{z}^j + \mathbf{N}\mathbf{Y}_T(\mathbf{z}^j) - \mathbf{k}] \tag{4–13}$$

converges to the solution $\mathbf{z}^*$ of Eq. 4–12 in such a way that

$$\|\mathbf{z}^* - \mathbf{z}^j\|_\infty \leq a^j\|\mathbf{z}^* - \mathbf{z}^0\|_\infty + O(h^p), \qquad a < 1$$

and

$$\|\mathbf{Y}_i(\mathbf{z}^j) - \mathbf{y}(t_i)\|_\infty \leq O(h^p) + a^j\|\mathbf{z}^* - \mathbf{z}_0\|_\infty,$$

where $\mathbf{y}(t)$ is the solution to the boundary value problem.

It often happens, however, that $1 - a$ is very small, so that the sequence $\mathbf{z}^j$ generated in Eq. 4–13 converges to $\mathbf{z}^*$ very slowly. By means of the connection matrix $\mathbf{C}(t;\, \mathbf{z})$ introduced in Section 2.5 it is possible to compute a more rapidly convergent sequence of $\mathbf{z}^j$. Recall that the n-by-n matrix $\mathbf{C}(t; \mathbf{z}) = [\partial \mathbf{y}(t; \mathbf{z})]/\partial \mathbf{z}$ satisfies

$$\frac{d}{dt}\mathbf{C}(t; \mathbf{z}) = \mathbf{J}(t, \mathbf{y}(t, \mathbf{z}))\mathbf{C}(t; \mathbf{z}), \tag{4–14}$$

where $\mathbf{C}(0; \mathbf{z}) = \mathbf{I}$ is the identity matrix and $\mathbf{J}(t, \mathbf{y}) = \partial \mathbf{f}(t, \mathbf{y})/\partial \mathbf{y}$. If we knew $\mathbf{C}(T; \mathbf{z})$, then we could estimate $\mathbf{z}^*$ via

$$\begin{aligned}\mathbf{k} &= \mathbf{M}\mathbf{z}^* + \mathbf{N}\mathbf{y}(T; \mathbf{z}^*)\\ &= \mathbf{M}(\mathbf{z} + \mathbf{z}^* - \mathbf{z}) + \mathbf{N}[\mathbf{y}(T; \mathbf{z}) + \mathbf{C}(T; \mathbf{z})(\mathbf{z}^* - \mathbf{z})] + O(\|\mathbf{z}^* - \mathbf{z}\|^2),\end{aligned}$$

so that, given $\mathbf{z} = \mathbf{z}^j$, we can compute $\mathbf{z}^{j+1}$ as

$$\mathbf{z}^{j+1} = \mathbf{z}^j + [\mathbf{M} + \mathbf{N}\mathbf{C}(T; \mathbf{z}^j)]^{-1}[\mathbf{k} - \mathbf{M}\mathbf{z}^j - \mathbf{N}\mathbf{y}(T; \mathbf{z}^j)];$$

this is Newton's method. Since it is known that $\mathbf{M} + \mathbf{N}\mathbf{C}$ is nonsingular, the iteration is defined. Computationally we would determine $\mathbf{C}(T; \mathbf{z}^j)$ as a numerical solution to Eq. 4–14 where we replace $\mathbf{J}(t, \mathbf{y}(t; \mathbf{z}^j))$ at $t = t_i$ with $\mathbf{J}(t_i, \mathbf{Y}_i(\mathbf{z}^j))$. This process usually converges rapidly to the sought solution $\mathbf{z}^*$.

Because we are solving our equation as an initial value problem, we may have introduced some stability difficulties. For example, the differential equation alone may have general solutions of the form e^{-t^2} and e^{t^2}, but the solution of the boundary value problem may be e^{-t^2}; shooting methods will always generate approximate solutions containing some portion of the e^{t^2} solution and will therefore cause some stability problems. In some cases this can be circumvented by integrating *backwards* in time; another somewhat different approach is to use *parallel* shooting.

Essentially, in parallel shooting, one divides the interval $[0, T]$ into subintervals $[T_i, T_{i+1}]$, $0 = T_0 < T_1 < \ldots < T_k = T$ and considers the differential equation independently over each subinterval, choosing initial conditions at $t = T_i$ and integrating to $t = T_{i+1}$; the various values at $t = T_i$, $i = 0, \ldots, K$ are then *simultaneously* adjusted to try to satisfy boundary conditions at 0 and T and continuity conditions at T_i, $1 \leq i \leq K - 1$. In practice, this appears to eliminate stability problems in many cases.

One of the problems in using shooting methods is that of obtaining a good first estimate of the entire initial vector; for this reason, and also because in many practical problems the correct initial vector is the desired output from a solution of a boundary value problem, we shall show how one can, in some cases, compute that initial vector or a good approximation to it. We consider a particular example to illustrate the technique:

$$y'' = t^2 + y^2, \qquad y(0) = y(1) = 0.$$

From Section 3.5 we know that we can consider this as an integral equation

$$g(t) = \int_0^1 K(t, \tau)[\tau^2 + y(\tau)^2]d\tau,$$

where $K(t, \tau)$ is a known positive piecewise-linear Green's function. By using the ideas of *interval analysis* (see Section 5.8 for a brief description of interval arithmetic) we can compute a sequence of interval-valued functions $Y_k(t)$ defined by

$$Y_{k+1}(t) \equiv \int_0^1 K(t, \tau)[\tau^2 + Y_k(\tau)^2]d\tau$$

$$Y_0(t) \equiv [-1, 0].$$

For example, the intervals $Y_1(t)$ and $Y_2(t)$ are

$$Y_1(t) = \left[-\frac{7t}{12} + \frac{t^2}{2} + \frac{t^4}{12}, -\frac{t}{12} + \frac{t^4}{12}\right]$$

$$Y_2(t) = \left[-\frac{193t}{1728} + \ldots, -\frac{145t}{1728} + \ldots\right].$$

It can be proved [82] by monotonicity arguments that

$$y(t) \in Y_k(t), \qquad k = 0, 1, \ldots, \qquad 0 \leq t \leq 1.$$

For $k \geq 1$ the intervals $Y_k(0)$ collapse to a single point set, namely

$$Y_k(0) = [0, 0], \qquad k \geq 1.$$

It follows that $y'(0)$ is contained in the interval of the coefficients of the linear part of $Y_k(t)$.

For $k = 1, 2$ we deduce that

$$-\tfrac{7}{12} \leq y'(0) \leq -\tfrac{1}{12} \qquad \text{for } k = 1,$$

$$-\tfrac{193}{1728} \leq y'(0) \leq -\tfrac{145}{1728} \qquad \text{for } k = 2.$$

It can be shown that $y'(0)$ can be computed to arbitrarily great accuracy in this fashion, since the widths of the intervals containing $y'(0)$ decrease by a factor of about 1/72 per iteration. For a more complex problem it might be too difficult to compute $y'(0)$ in this way, but this method certainly is useful for providing an estimate of $y'(0)$ for use in a shooting method.

Additional references: 45, 64.

4.11 INTEGRAL-EQUATION METHODS FOR BOUNDARY VALUE PROBLEMS

In Section 3.5 we demonstrated how boundary value problems can often be reformulated as integral equations; this provides a tool for computation, since numerical methods for solving integral equations have been widely studied. As an example of the ideas involved, we look again at the equation studied in Section 3.5,

$$\mathbf{y}'' = f(t, \mathbf{y}), \qquad \mathbf{y} \in E^n, \qquad \mathbf{y}(0) = \mathbf{y}_0, \qquad \mathbf{y}(T) = \mathbf{y}_T, \tag{4–15}$$

which we reduced to the equivalent integral equation

$$\mathbf{y}(t) = \mathbf{y}_0 + (\mathbf{y}_T - \mathbf{y}_0)\frac{t}{T} + \int_0^T K(t, \tau)\mathbf{f}(\tau, \mathbf{y}(\tau))d\tau.$$

If we set

$$\mathbf{w}(t) = \mathbf{y}(t) - \mathbf{y}_0 - (\mathbf{y}_T - \mathbf{y}_0)\frac{t}{T},$$

$$\mathbf{g}(t, \mathbf{w}) = \mathbf{f}\left(t, \mathbf{w} + \mathbf{y}_0 + (\mathbf{y}_T - \mathbf{y}_0)\frac{t}{T}\right),$$

then the integral equation becomes

$$\mathbf{w}(t) = \int_0^T K(t, \tau)\mathbf{g}(\tau, \mathbf{w}(\tau))d\tau. \tag{4–16}$$

For notational convenience we define the operator $\mathbf{G}$ and the integral operator $\mathbf{K}$ by

$$\mathbf{G}\mathbf{y} = \mathbf{g}(t, \mathbf{y}(t))$$

$$\mathbf{K}\mathbf{y} = \int_0^T K(t, \tau)\mathbf{y}(\tau)d\tau$$

so that Eq. 4–16 can be written merely as

$$\mathbf{w} = \mathbf{K}\mathbf{G}\mathbf{w}.$$

For a continuous function $\mathbf{w}(t)$ on $0 \leq t \leq T$, define

$$\|\mathbf{w}\|_T = \max_{0 \leq t \leq T} \|\mathbf{w}(t)\|_\infty.$$

If $\mathbf{g}(t, \mathbf{y})$ satisfies a Lipschitz condition with Lipschitz constant L_1 independent of t and $\mathbf{y}$ and if $\int_0^T |K(t, \tau)| d\tau \leq L_2$ for $0 \leq t \leq T$, then

$$\|\mathbf{KGu} - \mathbf{KGv}\|_T \leq L_1 L_2 \|\mathbf{u} - \mathbf{v}\|_T.$$

If $L_1L_2 < 1$ (so that the mapping KG is a *contraction*), it is possible to show, as was indicated in Sections 2.2 and 2.3, that the Picard-iteration method

$$\mathbf{w}^{j+1} = \mathbf{KGw}^j$$

yields a sequence of functions $\mathbf{w}^j(t)$ converging uniformly to the unique solution $\mathbf{w}(t)$ of Eq. 4–16 and thus yielding the solution $\mathbf{y}(t)$ of Eq. 4–15.

EXERCISE

Prove the uniform convergence of the Picard iterates to the solution.

A more rapidly convergent process can often be obtained by using Newton's method of local linearization. Let

$$\mathbf{J}(t, \mathbf{w}) = \frac{\partial \mathbf{g}(t, \mathbf{w})}{\partial \mathbf{w}}$$

exist, and define the operator $\mathbf{G}'_{\mathbf{w}}$ by

$$\mathbf{G}'_{\mathbf{w}}\mathbf{y} = \mathbf{J}(t, \mathbf{w}(t))\mathbf{y}(t).$$

If $\mathbf{w}(t)$ solves Eq. 4–16, then, given any other function $\mathbf{w}_0(t)$, we define

$$\mathbf{d}(t) = \mathbf{w}(t) - \mathbf{w}_0(t)$$

and write

$$0 = \mathbf{w}(t) - \int_0^T K(t, \tau)\mathbf{g}(\tau, \mathbf{w}(\tau))d\tau$$

$$0 = \mathbf{w}_0(t) + \mathbf{d}(t) - \int_0^T K(\mathbf{t}, \tau)[\mathbf{g}(\tau, \mathbf{w}_0(\tau)) + \mathbf{J}(\tau, \mathbf{w}_0(\tau))d(\tau)]d\tau + O(\|d\|_T^2).$$

Therefore we can approximate $\mathbf{d}(t)$ by $\mathbf{d}_0(t)$ where $\mathbf{d}_0(t)$ solves the *linear* integral equation

$$\mathbf{d}_0(t) - \int_0^T K(t, \tau)\mathbf{J}(\tau, \mathbf{w}_0(\tau))\mathbf{d}_0(\tau)d\tau = \mathbf{w}_0(t) - \int_0^T K(t, \tau)\mathbf{g}(\tau, \mathbf{w}_0(\tau))d\tau,$$

or, in compact notation,

$$\mathbf{d}_0 - \mathbf{K}\mathbf{g}'_{\mathbf{w}_0}\mathbf{d}_0 = \mathbf{w}_0 - \mathbf{KGw}_0.$$

If we then define

$$\mathbf{w}_1(t) = \mathbf{w}_0(t) + \mathbf{d}_0(t),$$

we can consider $\mathbf{w}_1(t)$ as a new approximation to $\mathbf{w}(t)$ and continue in this way. Thus

$$\mathbf{w}_{i+1}(t) = \mathbf{w}_i(t) + \mathbf{d}_i(t),$$

where $\mathbf{d}_i(t)$ solves

$$\mathbf{d}_i - \mathbf{K}\mathbf{G}'_{\mathbf{w}_i}\mathbf{d}_i = \mathbf{w}_i - \mathbf{K}\mathbf{G}\mathbf{w}_i.$$

This iteration is Newton's method; the asymptotic nature of the convergence is such that

$$\|\mathbf{w} - \mathbf{w}_{i+1}\|_T = O(\|\mathbf{w} - \mathbf{w}_i\|_T^2).$$

We remark that the Picard-iteration process requires evaluation of integral operators; this is usually done by a quadrature formula

$$\int_0^T \mathbf{w}(t)dt \sim \sum_{i=1}^m \alpha_i \mathbf{w}(t_i).$$

Newton's method, however, requires solution of a linear integral equation at each step. This commonly is accomplished by replacing the integral equation by a finite system of equations via a quadrature rule; that is,

$$\mathbf{u}(t) - \int_0^T P(t, \tau)\mathbf{u}(\tau)d\tau = \mathbf{r}(t)$$

is solved for the unknowns $u(t_i)$ satisfying

$$\mathbf{u}(t_i) - \sum_{j=1}^m A_j P(t_i, \tau_j)\mathbf{u}(t_j) = \mathbf{r}(t_i).$$

This approach can be used directly on the nonlinear integral equation Eq. 4–16 but leads to a finite system of nonlinear "algebraic" equations which can be solved approximately, using iterative methods such as Newton's method. Further details of the properties of these procedures are beyond the scope of this book.

EXERCISE

Replacing the integral $\int_0^1 s(t)\, dt$ by the trapezoidal quadrature sum $\{\frac{1}{2}s(0) + \frac{1}{2}s(1) + \sum_{i=1}^{N-1} s(ih)\}\, h$ where $Nh = 1$, find the system of equations resulting from thus discretizing Eq. 4–16 applied to solve $y'' = e^y$, $y(0) = 1$, $y(1) = 2$.

Additional references: 3, 45, 64, 84, 86.

4.12 ACCELERATION METHODS

In most of the numerical methods we have discussed in this chapter, we have envisioned computing different approximate solutions converging to a desired solution for some parameter h tending to zero; thus we might hope to be able to accelerate the convergence of this process.

In general—although we have not presented enough material to indicate this—we not only have asymptotic bounds on the numerical solution of the form

$$|\mathbf{y} - \mathbf{Y}| = O(h^p)$$

but we also have results on asymptotic behavior of the form

$$\mathbf{Y}(h) = \mathbf{y} + \mathbf{a}_0 h^{p_0} + \mathbf{a}_1 h^{p_1} + \ldots, \qquad p = p_0 < p_1 < \ldots$$

where by $\mathbf{Y}(h)$ we mean the entire numerical solution obtained with the parameter h; the parameters $\mathbf{a}_0, \mathbf{a}_1, \ldots$ are *unknown.* If we next compute $\mathbf{Y}(\lambda h)$ for $\lambda < 1$, then

$$\frac{1}{1 - \lambda^{p_0}}[\mathbf{Y}(\lambda h) - \mathbf{Y}(h)] = \mathbf{y} + \mathbf{b}_1 h^{p_1} + \ldots,$$

and thus we have computed an $O(h^{p_1})$ solution from two $O(h^{p_0})$ solutions. This acceleration method is called *extrapolation to the limit* and can be used recursively to compute solutions accurate to $O(h^{p_2}), \ldots$. At some point, of course, perhaps frustratingly soon, rounding errors will obliterate the desired acceleration.

EXERCISE

Consider the application of Euler's method to solve $\mathbf{y}' = \mathbf{f}(\mathbf{y})$, $\mathbf{y}(0) = \mathbf{y}_0$ having exact solution $\mathbf{y}(t)$. *Assume* that the resulting error $\mathbf{Y}_i - \mathbf{y}(ih) \equiv \mathbf{E}_i$ can be written $\mathbf{E}_i = \mathbf{E}(ih)h + O(h^2)$ for a function $\mathbf{E}(t)$. Try to see why $\mathbf{E}(t)$ satisfies the differential equation

$$\mathbf{E}' = \mathbf{f}_y(\mathbf{y}(t))\mathbf{E} - \tfrac{1}{2}\mathbf{y}''(t), \ \mathbf{E}(0) = \mathbf{0}.$$

Hint: $(1/h)\mathbf{E}_i$ satisfies a difference equation that is a negligible perturbation of the Euler's-method equations derived from the differential equation for $\mathbf{E}(t)$.

Another method which is in a sense an acceleration method is the *deferred correction* or *difference correction* method. In this approach we compute higher-and-higher-order accuracy solutions not by changing h but by changing the method in a simple manner. For example, consider solving the initial value problem

$$\mathbf{y}' = \mathbf{f}(\mathbf{y}), \qquad \mathbf{y}(0) = \mathbf{y}_0$$

by the first-order explicit formula

$$\mathbf{Y}_{i+1}^0 = \mathbf{Y}_i^0 + h\mathbf{f}(t_i, \mathbf{Y}_i^0);$$

this gives a solution $\mathbf{Y}_i^0$ satisfying

$$\mathbf{Y}_i^0 = \mathbf{y}(t_i) + O(h).$$

A higher-order extension of this formula is the second-order Taylor formula

$$\mathbf{Y}_{i+1} = \mathbf{Y}_i + h\mathbf{f}(t_0, \mathbf{Y}_i) + \frac{h^2}{2}\left[\frac{d}{dt}\,\mathbf{f}(t, \mathbf{Y}(t))\right]_{t = t_i,\ \mathbf{Y} = \mathbf{Y}_i}$$

We replace the latter derivative by $(1/h)[\mathbf{f}(t_{i+1}, \mathbf{Y}_{i+1}^0) - \mathbf{f}(t_i, \mathbf{Y}_i^0)]$ and then solve for the new solution $\mathbf{Y}_i^1$ from the explicit formula

$$\mathbf{Y}_{i+1}^1 = \mathbf{Y}_i^1 + h\mathbf{f}(t_i, \mathbf{Y}_i^1) + \frac{h}{2}\,[\mathbf{f}(t_{i+1}, \mathbf{Y}_{i+1}^0) - \mathbf{f}(t_i, \mathbf{Y}_i^0)].$$

It can be shown [37] that

$$\mathbf{Y}_i^1 = \mathbf{y}(t_i) + O(h^2)$$

and, if one proceeds further in this fashion,

$$\mathbf{Y}_i^j = \mathbf{y}(t_i) + O(h^{j+1}), \qquad j = 0, 1, \ldots;$$

computer programs are available [37] to generate the numerical differentiation formulas needed in the further terms of the expansion. Precisely the same approach can be used on difference methods for boundary value problems. This method has been thoroughly developed for two-point boundary value problems and appears on the basis of numerical experimentation to be one of the best techniques for obtaining solutions of high accuracy [45, 73, 87]; the application to initial value problems is more recent and has not been as thoroughly studied or tested.

EXAMPLE In Section 4.9, we considered the standard fourth-order difference method applied to solve

$$y'' = y^3 - (1 + \sin^2 t)\sin t, \qquad y(0) = y(\pi) = 0$$

with step sizes of $h = \pi/10$ and $\pi/20$, for which the maximum errors were 1.2×10^{-5} and 7.4×10^{-7} respectively. If instead of changing the step size one uses deferred corrections starting with the 1.2×10^{-5} accurate solution, one can easily compute [87] successive solutions with maximum errors of 4.2×10^{-9}, 2.2×10^{-12}, 3.2×10^{-15}, and 1.8×10^{-17}. ●

Additional references: 14, 37, 44, 45, 50, 64, 73, 87, 88.

4.13 METHODS FOR PERIODIC SOLUTIONS

As we stated in Section 2.10, the theory of periodic solutions of differential equations is somewhat less developed than the theory of either initial or boundary value problems; this is also true of numerical methods. We give here some indications of the types of approaches available.

Suppose one seeks a periodic solution of unknown period T to the autonomous system

$$\frac{d\mathbf{y}}{dt} = \mathbf{f}(\mathbf{y}), \qquad \mathbf{y} \in E^n.$$

Introduce the new variable $\tau = t/T$ and define

$$\mathbf{z}(\tau) = \mathbf{y}(t) = \mathbf{y}(\tau T).$$

Then

$$\frac{d\mathbf{z}}{d\tau} = T\mathbf{f}(\mathbf{z});$$

we seek a solution $\mathbf{z}(\tau)$ of period one. This can be treated by a shooting method. Select initial data $\mathbf{z}_0$ and a trial period T_0; integrate the initial value problem

$$\frac{d\mathbf{z}}{d\tau} = T_0\mathbf{f}(\mathbf{z}), \qquad \mathbf{z}(0) = \mathbf{z}_0$$

for $0 \le \tau \le 1$. Finally, use Newton's method or some other technique to adjust $\mathbf{z}_0$ and T_0 so that $\mathbf{z}(1)$ will be nearer to $\mathbf{z}(0) = \mathbf{z}_0$. A modified special version of this technique, but still in essence a shooting method using Newton's method, has been applied [102] to compute periodic solutions to the van der Pol equation

$$x'' - \lambda(1 - x^2)x' + x = 0$$

or, in system form,

$$x' = y$$

$$y' = -x + \lambda(1 - x^2)y,$$

for values of the parameter λ from 0 to 10. Newton's method works well here and on most such problems, since generally an approximate solution is known from perturbation methods. Numerical results obtained in this way appear to be accurate, although exact solutions are unknown.

EXERCISE

Consider finding the unknown period and a periodic solution of $y_1' = y_2$, $y_2' = -y_1$ by the above shooting method. Let T be the unknown period, change variables to change the period to equal one, and solve the resulting system exactly with initial data z_0. Solve for z_0 and T_0 in the equations stating that the solution has period equal to one, thus obtaining the period and a solution.

In cases in which the period is known, one can still, of course, use the method above or perhaps some boundary value method; another method of value is the following. Consider the time-dependent system

$$\mathbf{y}' = f(t, \mathbf{y}), \qquad \mathbf{y} \in E^n,$$

where $\mathbf{f}(t, \mathbf{y})$ is periodic of period, say, 2π in t, and we seek a solution with that period. We approximate $\mathbf{y}(t)$ by a truncated Fourier series

$$\mathbf{Y}_k(t) = \tfrac{1}{2}\mathbf{a}_0 + \sum_{j=1}^{k}(\mathbf{a}_{2j-1} \sin jt + \mathbf{a}_{2j} \cos jt)$$

and seek to determine the vectors $\mathbf{a}_0, \mathbf{a}_1, \ldots, \mathbf{a}_{2k} \in E^n$. Now $\mathbf{f}(t, \mathbf{Y}_k(t))$ also has a truncated Fourier expansion

$$\mathbf{f}(t, \mathbf{Y}_k(t)) = \tfrac{1}{2}\mathbf{b}_0 + \sum_{j=1}^{k}(\mathbf{b}_{2j-1} \sin jt + \mathbf{b}_{2j} \cos jt)$$

where

$$\mathbf{b}_{2j} = \frac{1}{\pi}\int_0^{2\pi} \mathbf{f}(s, \mathbf{Y}_k(s)) \cos js \, ds$$

$$\mathbf{b}_{2j-1} = \frac{1}{\pi}\int_0^{2\pi} \mathbf{f}(s, \mathbf{Y}_k(s)) \sin js \, ds;$$

the vectors $\mathbf{b}_j = \mathbf{b}_j(\mathbf{a}_0, \ldots, \mathbf{a}_{2k})$ are functions of $\mathbf{a}_0, \ldots, \mathbf{a}_{2k}$. Replacing $\mathbf{y}$ and $\mathbf{f}(t, \mathbf{y})$ by their truncated series in the differential equation and equating terms yields the following system of equations for the vectors $\mathbf{a}_0, \ldots, \mathbf{a}_{2k}$:

$$\mathbf{b}_0(\mathbf{a}_0, \mathbf{a}_1, \ldots, \mathbf{a}_{2k}) = 0$$

$$\mathbf{b}_{2j-1}(\mathbf{a}_0, \mathbf{a}_1, \ldots, \mathbf{a}_{2k}) + j\mathbf{a}_{2j} = 0, \qquad j = 1, \ldots, k,$$

$$\mathbf{b}_{2j}(\mathbf{a}_0, \mathbf{a}_1, \ldots, \mathbf{a}_{2k}) - j\mathbf{a}_{2j-1} = 0, \qquad j = 1, \ldots, k.$$

The trigonometric polynomial $\mathbf{Y}_k(t)$ corresponding to a real solution of this system is taken as an approximation to the true solution $\mathbf{y}(t)$. If $\mathbf{f}(t, \mathbf{y})$ is continuously differentiable in t and $\mathbf{y}$, and if $\mathbf{y}(t)$ is an isolated periodic solution, then for large k we can compute approximations $\mathbf{Y}_k(t)$ such that $\mathbf{Y}_k$ and $\mathbf{Y}_k'$ converge uniformly to $\mathbf{y}$ and $\mathbf{y}'$ [101]. Under certain circumstances, error bounds can be computed and the *existence* of $\mathbf{y}$ can be asserted from computational results.

EXAMPLE [101] Consider the equation

$$x'' + x^3 = \sin t$$

or, in system form,

$$x' = y$$

$$y' = -x^3 + \sin t,$$

for which a solution of period 2π is known to exist. If we take $k = 15$ and use the *Galerkin method* described above, it can be shown that there exists precisely one isolated periodic solution $x(t)$ near our approximation $x_{15}(t)$ satisfying

$$|x_{15}(t) - x(t)| \leq 0.0000083. \quad \bullet$$

Additional reference: 100.

5

SPECIAL NUMERICAL METHODS FOR INITIAL VALUE PROBLEMS

5.1 INTRODUCTION

In this chapter we wish to discuss certain methods for solving initial value problems; in general these methods are less straightforward in conception or are less widely used than those of the preceding chapter. In jumping from the standard basic methods of Chapter 4 to those we are about to discuss, we leave out a large number of special, recently developed methods that are of the *type* discussed already but are designed to exhibit some special attractive properties. Much fruitful work has been performed by skillful analysts to devise specific one-step and linear multistep methods that are highly accurate, stable, and convenient to use computationally; however, it is not possible to discuss them all. Although each of these methods has its special characteristics, most of them fit into the general ideas already presented; therefore we reluctantly choose to mention none specifically. Rather, we continue in this chapter and the next to look at computational methods in a general way, delineating broad classes of methods and interesting new approaches.

Additional references: 15, 16, 41, 42, 69, 70, 71, 72, 83, 91, 94, 95, 103, 109.

5.2 METHODS OF BEST FIT

A number of interesting methods are based on the idea of finding the best fit, or best approximation, in some sense, to the solution by using a given set of functions. We describe briefly how some of these techniques can be applied to solve

$$\mathbf{y}' = \mathbf{f}(t, \mathbf{y}), \qquad \mathbf{y} \in E^n, \qquad 0 \leq t \leq T \tag{5–1}$$

with the initial condition

$$\mathbf{y}(0) = \mathbf{y}_0. \tag{5–2}$$

Consider a collection of J functions $\varphi_j(t) \in E^n$; for a vector $\boldsymbol{\alpha} \in E^J$, define

$$\boldsymbol{\phi}(t; \boldsymbol{\alpha}) = \sum_{j=1}^{J} \alpha_j \varphi_j(t).$$

Discretize the interval $[0, T]$ by a grid $0 = t_0 < t_1 < \ldots < t_N = T$. One method of best fit is to choose $\boldsymbol{\alpha}$ such that

$$\|\boldsymbol{\phi}'(t; \boldsymbol{\alpha}) - \mathbf{f}(t, \boldsymbol{\phi}(t; \boldsymbol{\alpha}))\|_N$$

is minimized over the class of $\boldsymbol{\alpha}$ satisfying $\boldsymbol{\phi}(0; \boldsymbol{\alpha}) = \mathbf{y}_0$; here $\|\mathbf{w}(t)\|_N$ is some nonnegative function (a semi-norm) based on the values $\|\mathbf{w}(t_i)\|$ above, for example, $\|\mathbf{w}(t)\|_N = \max_{0 \leq i \leq N} \|\mathbf{w}(t_i)\|$. If $J \geq (N + 2)n$, then in general one can find $\boldsymbol{\alpha}$ such that

$$\|\boldsymbol{\phi}'(t; \boldsymbol{\alpha}) - \mathbf{f}(t, \boldsymbol{\phi}(t; \boldsymbol{\alpha}))\|_N = 0;$$

this is the *collocation* method. If $(N + 2)n > J$, then we have a nonlinear problem in mathematical programming in order to solve for $\boldsymbol{\alpha}$. A slight modification of this is to include $\|\boldsymbol{\phi}(0; \boldsymbol{\alpha}) - \mathbf{y}_0\|$ in the function $\|\cdot\|_N$ and not restrict $\boldsymbol{\alpha}$ at all.

EXAMPLE Consider the simple differential equation

$$y' = y, \qquad y \in E^1, \qquad 0 \leq t \leq 1, \qquad y(0) = 1$$

having solution $y(t) = e^t$. We seek a good approximation of the form $\phi(t) = a_1 + a_2 t$. The initial condition requires $a_1 = 1$. We seek to minimize

$$\|\phi' - \phi\|_N = \max_{0 \leq i \leq N} |a_2 - (1 + a_2 t_i)|, \qquad t_i \text{ equally spaced.}$$

If $N = 0$, the collocation method gives $\phi(t) = 1 + t$. For $N > 0$, we have a linear programming problem, the solution to which is $a_2 = 1$ for all N, so we set $\phi(t) = 1 + t$. The maximum of the error $|e^t - (1 + t)|$ is roughly 0.72. Using parabolas $\phi(t) = a_1 + a_2 t + a_3 t^2$ similarly we get an error of roughly 0.15. ●

EXERCISE

Use a computer and a linear programming routine to approximate on [0, 1] the solution $\sin t$ to $y'' + y = 0$, $y(0) = 0$, $y'(0) = 1$ by polynomials of degrees 1, 2, 3, 4, and 5 as described above.

In the description above we did not indicate what is done when $(N + 2)n < J$; in this case one tries to make the solution as smooth as possible in some sense. Rather than describe this general method precisely, we give an example.

EXAMPLE Consider again the equation of the last example. Let us approximate the solution for $0 \leq t \leq \frac{1}{2}$ by a function $a + bt + ct^2$ and, for the interval $\frac{1}{2} \leq t \leq 1$, by a function $d + ft + gt^2$. We shall choose a, b, c, d, f, and g so as to satisfy the initial condition; to satisfy the differential equation at $t = 0, \frac{1}{2}$, and 1; and so that the approximate solution has as many continuous derivatives at $t = \frac{1}{2}$ as possible. This leads us to an approximate solution by the *spline function*

$$\phi(t) = \begin{cases} 1 + t + \frac{2}{3}t^2 & \text{for } 0 \leq t \leq \frac{1}{2} \\ \frac{10}{9} + \frac{5}{9}t + \frac{10}{9}t^2 & \text{for } \frac{1}{2} \leq t \leq 1. \end{cases}$$

$\phi(t)$ is continuously differentiable on [0, 1]; $\phi''(t)$ equals 4/3 and 20/9, respectively, on the two intervals and hence is not continuous. The maximum error $|\phi(t) - e^t| = |\phi(1) - e^1| \leq 0.06$. ●

Under reasonable hypotheses all these best-fit methods can be shown to converge as J and N grow.

Additional references: 68, 89.

5.3 BOUNDARY VALUE METHODS

Questions of stability are perhaps most important in the solution of initial value problems over a semi-infinite interval, say $0 \leq t \leq \infty$. As we have mentioned before, it is sometimes true that although the *general* solution of a given equation may not tend to zero as t tends to infinity, the *particular* solution (or some of its components) determined by the given initial data may be known to converge to zero, either for physical or theoretical reasons. In such a case or in any situation in which we have knowledge of the precise asymptotic behavior of a desired solution, it is sometimes useful to consider the problem as a boundary value problem with some such condition as $\mathbf{y}(\infty) = \mathbf{0}$ imposed in addition to some components of the initial condition $\mathbf{y}(0) = \mathbf{y}_0$; in practice, of course, this requires setting $\mathbf{y}(T) = \mathbf{0}$ for some large T. The extra condition forces the solution to stay near zero for large $t \leq T$ and thereby appears to eliminate the effect of nondecaying terms.

EXAMPLE Consider the *Airy equation*

$$y'' = ty, \qquad y \in E^1. \tag{5–3}$$

This equation has two independent solutions denoted by $Ai(t)$ and $Bi(t)$; it is known that $Ai(t) > 0$, $\lim Ai(t) = 0$, $\lim Bi(t) = \infty$. The initial conditions

$$y(0) = 0.355028053887817\ldots \tag{5–4}$$

$$y'(0) = -0.258819403792807\ldots \tag{5–5}$$

determine the function $Ai(t)$ as the solution [1]; because of the presence of $Bi(t)$ it is very difficult to compute $Ai(t)$ accurately from the differential equation by standard initial value methods. The solution has been computed [52] by solving Eq. 5–3 with the boundary condition in Eq. 5–4 and with $y(10\pi) = 0$; using the formula

$$Y_{i+1} - 2Y_i + Y_{i-1} = h^2 t_i Y_i$$

with a step size of $h = (\pi/20) \sim .16$, we find that the numerical solution obtained is accurate to at least three decimal places. ●

In general it is dangerous to replace an initial value problem by a boundary value problem since they need not be equivalent. The equation

$$y'' = -y, \qquad y(0) = 0, \qquad y'(0) = 1$$

has the unique solution $y(t) = \sin t$ which is periodic of period 2π; the boundary value problem

$$y'' = -y, \qquad y(0) = y(2\pi) = 0,$$

however, has infinitely many solutions, $y = \alpha \sin t$ for any α. For the example using the Airy function, the boundary problem with $y(\infty) = 0$ is known to be equivalent to the initial value problem. One must exercise care in using this approach, but when applicable it appears to be useful. The method can be used equally well on partial differential equations and in fact may be of greater importance there. For parabolic partial differential equations the theory has been investigated by A. Carasso and S. Parter in University of Wisconsin Computer Sciences Technical Report #39.

5.4 SUPER-STABLE METHODS

We mentioned in Section 2.2 and at the conclusion of Section 4.7 that one needs methods with very strong stability properties for linear equations with eigenvalues having negative real parts of which at least one is much larger in modulus than some other one; it would be of value to have a super-stable numerical method giving a numerical solution guaranteed to converge to zero *independent of the size of h*, the discretization parameter. By analogy we would expect such methods to be of value for nonlinear equations also. In recent years much effort has gone into the development of such methods or similar ones designed to treat the same kind of problem. For clarity, we consider the equation

$$y' = \lambda y, \qquad y \in E^1, \qquad y(0) = 1, \qquad \lambda \text{ complex},$$

where the real part of λ, $Re(\lambda)$, is negative. We seek to solve this by some kind of discrete-variable method, computing an approximate solution Y_i at $t_i = ih$; we desire the method to be such that $\lim_{i\to\infty} Y_i = 0$. A number of such super-stable methods have been developed which guarantee that this property will hold whenever λh lies in a certain region of the complex plane containing regions in which $|\lambda h|$ is large. For simplicity, we look briefly at methods for which Y_i tends to zero *for all* λh with $Re(\lambda h) < 0$; such methods are called *A-stable.*

It has been shown [31, 34] that no explicit type of formula can be A-stable; for example, we saw in Section 4.7 that the Runge-Kutta-method solution tended to zero only for h small. Dahlquist [31, 34] has analyzed the general linear multistep method described in Section 4.5,

$$\alpha_k Y_{i+1} + \ldots + \alpha_0 Y_{i+1-k} = \lambda h(\beta_k Y_{i+1} + \ldots + \beta_0 Y_{i+1-k}),$$

and has shown that no A-stable method can be of order greater than two. If one defines the *error constant C* as the number such that

$$Y = y + Ch^p + O(h^{p+1}),$$

then it is known that of all A-stable linear multistep methods of order two, the smallest error constant is exhibited by the *trapezoidal rule*

$$Y_{i+1} = Y_i + \tfrac{1}{2}h(f_{i+1} + f_i).$$

EXERCISE

Verify the A-stability of the trapezoidal rule.

It is possible, however, to find other classes of A-stable methods. One can, for example, consider linear multistep methods in which higher derivatives than $y' = f(t, y)$ can appear, that is

$$\alpha_k Y_{i+1} + \ldots + \alpha_0 Y_{i+1-k} = \sum_{j=0}^{J} h^j(\beta_{jk} Y_{i+1}^{(j+1)} + \ldots + \beta_{j0} Y_{i+1-k}^{(j+1)}), \quad (5\text{–}6)$$

where

$$Y_i^{(j)} = \left[\frac{d^{j-1}}{dt^{j-1}} f(t, w(t))\right]_{t = t_i,\ w = Y_i}$$

A natural generalization of the trapezoidal rule to this setting is the formula

$$Y_{i+1} = Y_i + h\sum_{j=0}^{J} d_{J,j}[Y_i^{(j+1)} + (-1)^j Y_{i+1}^{(j+1)}]. \quad (5\text{–}7)$$

where the $d_{J,j}$ are known constants [37, 75, 76]; for $J = 0$ this becomes the trapezoidal rule. This method, using J formal derivatives of $f(t, y)$, is A-stable and of order $2J + 2$. Moreover, if L steps of the deferred-correction process discussed in Section 4.12 are applied to this method, the resulting method remains A-stable (this is valid for any A-stable original method) and is of order $(L + 1)(2J + 2)$. It is conjectured here that no A-stable method of the form of Eq. 5–6 can be of order greater than $2J + 2$ and that, of those A-stable methods of order $2J + 2$, the smallest error constant is exhibited by the *Hermite method* of Eq. 5–7.

One of the difficulties in using the class of methods defined by Eq. 5–7 to solve $y' = \lambda y$ for $Re(\lambda h) < 0$, $|\lambda h|$ large, lies in the fact that one must solve an implicit equation for Y_{i+1} at each step. The usual iterative method for solving this is known to converge for $|\lambda h|$ small, but we have $|\lambda h|$ large. One can usually circumvent this difficulty by using a modification of Newton's method to solve for Y_{i+1}; this problem does, however, present one of the drawbacks of A-stable methods and implicit methods in general.

These results demonstrate that, in a sense, the property of A-stability is almost too much to ask of a method; few techniques meet the standard. Often one can be satisfied if Y_i will tend to zero for λh in, say, a pie-shaped wedge about the origin of the complex plane or in some other region. Such methods are currently being developed and studied.

Additional references: 30, 31, 34, 48, 72, 85.

5.5 OTHER MULTISTEP METHODS

In the preceding section, Eq. 5–6 and Eq. 5–7 describe modifications of the standard linear multistep method described in Section 4.5; we shall briefly discuss some other modifications which have been developed recently and appear to be of value.

Consider the standard method

$$\alpha_k \mathbf{Y}_{i+1} + \ldots + \alpha_0 \mathbf{Y}_{i+1-k} = h(\beta_k \mathbf{f}_{i+1} + \ldots + \beta_0 \mathbf{f}_{i+1-k}), \qquad (5\text{–}8)$$

and let

$$\rho(z) = \alpha_k z^k + \ldots + \alpha_0$$

be of exact degree $k_\rho = k$, and

$$\sigma(z) = \beta_k z^k + \ldots + \beta_0$$

be of exact degree $k_\sigma \leq k$. It is known [32, 55] that, in order for the method of Eq. 5–8 to be D-stable, the order p of the method cannot exceed $k + 1$ if k is odd and $k + 2$ if k is even. Thus, although there are $k_\rho + k_\sigma + 1$ free parameters for the method, we cannot use them merely to get maximal order if we wish to maintain stability! Modifications in the method have been introduced to avoid this problem.

If $\beta_k \neq 0$, one normally implements Eq. 5–8 via a predictor-corrector scheme as in Section 4.6; given this fact, it is clear that one could predict $\mathbf{Y}_{i+1-r}$ for some *noninteger value* r just as easily as $\mathbf{Y}_{i+1}$ and could perhaps make good use of this value. This approach has been of recent interest, both with the value of r being given and with it being chosen to maximize the order while maintaining D-stability. It has been shown [51] that, for $k_\sigma \leq k_\rho = k \leq 4$, r can be chosen so that the method has order $p = k_\sigma + k_\rho + 2$, the maximal conveivable order obtainable by using $k_\sigma + k_\rho + 3$ free parameters; moreover, the method is D-stable. This means, for example, that with $k_\sigma + 1 = k_\rho = 4$ we get a D-stable method of order 9 requiring the same amount of computation as the optimal sixth-order method of Eq. 5–8 with $k_\rho = k_\sigma = 4$.

We remark that the deferred-correction method mentioned in Section 4.12 can be viewed, when applied to certain methods, as a modification of the standard multistep method allowing $k_\rho < k_\sigma$; this leads to the idea of such advanced multistep methods in general and may be of some value [35].

Although it actually fits into the standard linear multistep and predictor-corrector method setting, there is one other class of methods of computational interest that we wish to mention, namely one specifically designed to make use of the parallel-processing capability of some modern digital computers. Let the superscripts p and c denote "predicted" value and "corrected" value respectively, and consider the method

$$\mathbf{Y}_{i+1}^p = \mathbf{Y}_{i-1}^c + 2h\mathbf{f}_i^p$$

$$\mathbf{Y}_i^c = \mathbf{Y}_{i-1}^c + \frac{h}{2}(\mathbf{f}_i^p + \mathbf{f}_{i-1}^c)$$

EXERCISE

Verify that, when the method above is applied to equations of the simple form $\mathbf{y}' = \mathbf{A}\mathbf{y}$, where $\mathbf{A}$ is constant, it is **D**-stable and of second order.

The computations described by these two equations may be performed in parallel, that is simultaneously, since neither depends on the other! Therefore this **D**-stable second-order method could be used, were parallel processors available, with roughly one-half the computer time required by the standard second-order method

$$\mathbf{Y}_{i+1}^p = \mathbf{Y}_i^c + \frac{h}{2}(3\mathbf{f}_i^c - \mathbf{f}_{i-1}^c)$$

$$\mathbf{Y}_{i+1}^c = \mathbf{Y}_i^c + \frac{h}{2}(\mathbf{f}_{i+1}^p + \mathbf{f}_i^c)$$

which must be implemented serially. Higher-order methods using two or more processors are also known [79].

Additional references: 47, 49.

5.6 LIE-SERIES METHODS

In Section 2.4 we showed how the Lie series gave a formal solution to the equation

$$\mathbf{y}' = \mathbf{f}(\mathbf{y}), \qquad \mathbf{y} \in E^n, \qquad \mathbf{y}(0) = \mathbf{Y}_0$$

in terms of a perturbation expansion based on a known solution $\mathbf{y}(t)$ of

$$\bar{\mathbf{y}}' = \mathbf{g}(\mathbf{y}), \qquad \bar{\mathbf{y}} \in E^n, \qquad \bar{\mathbf{y}}(0) = \mathbf{y}_0$$

and in terms of differential operators **F**, **G**, and **D** defined there. This series can be used also to compute numerical approximations.

Suppose that we seek a solution $\mathbf{y}(t)$ first on the interval $0 \leq t \leq h$ and that we have found a $\bar{\mathbf{y}}(t)$ that satisfies

$$\bar{\mathbf{y}}(t) = \mathbf{y}(t) + O(h^m), \qquad 0 \leq t \leq h,$$

perhaps by letting $\bar{\mathbf{y}}(t)$ be the first m terms of the Taylor series for $\mathbf{y}(t)$ at $t = 0$. Suppose also that we truncate the Lie series at $(s + 1)$ terms to define the approximation $\hat{\mathbf{y}}(t)$, that is,

$$\hat{\mathbf{y}}(t) = \bar{\mathbf{y}}(t) + \sum_{j=0}^{s} \int_0^t \frac{(t - \tau)^j}{j!} [\mathbf{DF}^j\mathbf{y}]_{\mathbf{y}=\bar{\mathbf{y}}(\tau)} d\tau;$$

then it can be shown [67] that

$$\hat{\mathbf{y}}(t) = \mathbf{y}(t) + O(h^{m+s+1}), \qquad 0 \leq t \leq h;$$

thus we can use Lie series to boost the order of our approximation considerably. Applying the process over again at h, $2h$, $3h$, . . . , we get, overall, a solution that is of order $m + s$.

The most serious problems involved in implementing this are in the generation of the expressions $\mathbf{DF}^j$ and in evaluating the integrals in the truncated series. Computer programs have been written which recursively perform the symbolic differentiations necessary to obtain $\mathbf{DF}^j$, whenever $\mathbf{f}(\mathbf{y})$ is well-enough behaved, and which compute the integrals by Gaussian quadrature formulas of sufficient accuracy to maintain overall accuracy [66]. It is also possible to estimate error and monitor the step size with little added effort; the program has been used experimentally and yields very accurate results.

EXAMPLE The Lie-series method has been used [67] to solve numerically the difficult restricted three-body equations

$$y_1' = y_2$$

$$y_2' = y_1 + 2y_4 - \mu_1 \frac{y_1 + \mu}{[(y_1 + \mu)^2 + y_3^2]^{3/2}} - \frac{y_1 - \mu_1}{[(y_1 - \mu_1)^2 + y_3^2]^{3/2}}$$

$$y_3' = y_4$$

$$y_4' = y_3 - 2y_3 - \mu_1 \frac{y_3}{[(y_1 + \mu)^2 + y_3^2]^{3/2}} - \frac{y_3}{[(y_1 - \mu_1)^2 + y_3^2]^{3/2}}$$

where $\mu_1 = 1 - \mu$.
Using known initial data to start the computation of a periodic orbit, the Lie-series method gave a solution which came within 0.03×10^{-16} of meeting the initial data at the completion of a numerical "orbit." This solution was of greater accuracy than solutions obtained by other methods applied to the same problem [39, 46] and used less computer time. ●

Additional references: 53, 65.

5.7 INTEGRAL-EQUATION METHODS

An integral-equation formulation of the initial value problem was used in Section 2.3 to prove existence and uniqueness; as we indicated there, the Picard iteration is also of some use for numerical computation. This iteration is, however, rather slowly convergent; the local-linearization method of Newton, discussed in Section 4.11 with respect to boundary value problems, is usually more rapidly convergent, and we shall therefore discuss it.

We consider the integral equation

$$\mathbf{y}(t) = \mathbf{y}_0 + \int_0^t \mathbf{f}(\mathbf{y}(\tau))d\tau$$

equivalent to our initial value problem

$$\mathbf{y}' = \mathbf{f}(\mathbf{y}), \qquad \mathbf{y} \in E^n, \qquad \mathbf{y}(0) = \mathbf{y}_0.$$

Assuming that the matrix

$$\mathbf{J}(\mathbf{y}) = \frac{\partial \mathbf{f}}{\partial \mathbf{y}}$$

exists, given an approximate solution $\mathbf{y}_0(t)$, we linearize and thus seek $\mathbf{y}_1(t)$ to solve

$$\mathbf{y}_1(t) = \mathbf{y}_0 + \int_0^t [\mathbf{f}(\mathbf{y}_0(\tau)) + \mathbf{J}(\mathbf{y}_0(\tau))(\mathbf{y}_1(\tau) - \mathbf{y}_0(\tau))]d\tau.$$

Having found $\mathbf{y}_1(t)$, we replace $\mathbf{y}_0(t)$ by $\mathbf{y}_1(t)$ and proceed iteratively. Note that $\mathbf{y}_1(t)$ also satisfies

$$\mathbf{y}_1' = \mathbf{f}(\mathbf{y}_0) + \mathbf{J}(\mathbf{y}_0)(\mathbf{y}_1 - \mathbf{y}_0), \qquad \mathbf{y}_1(0) = \mathbf{y}_0(0).$$

EXAMPLE Consider the equation

$$y' = y^2, \qquad y \in E^1, \qquad y(0) = 1,$$

having solution $y(t) = 1/(1 - t)$. We have considered this problem before in Section 2.3 and Section 2.4. Here we have $J(y) = 2y$. If we let $y_0(t) \equiv y_0 = 1$, we find that $y_1(t)$ must satisfy

$$y_1' = 1 + 2[y_1 - 1]$$

which yields

$$y_1(t) = \tfrac{1}{2}e^{2t} + \tfrac{1}{2} = 1 + t + t^2 + \tfrac{2}{3}t^3 + \dots.$$

We then find that y_2 satisfies

$$y_2' = (e^{2t} + 1)y_2 - \tfrac{1}{4}(e^{2t} + 1)^2, \qquad y_2(0) = 1$$

which implies

$$y_2(t) = 1 + t + t^2 + t^3 + t^4 + t^5 + t^6 + \tfrac{62}{63}t^7 + \dots. \quad \bullet$$

Although the sequence $\mathbf{y}_i(t)$ seems to converge rapidly in general, its computation requires solution of a linear integral equation (or differential equation) at each step. To avoid this, we approximate $(\mathbf{I} - \mathbf{K})^{-1}$ by $\mathbf{I} + \mathbf{K}$, where $\mathbf{K}$ is the integral operator defined by the kernel of the integral equation, and then consider the following simpler iteration,

$$\mathbf{y}_1(t) = \mathbf{y}_{1/2}(t) + \int_0^t \mathbf{J}(\mathbf{y}_0(\tau))(\mathbf{y}_{1/2}(\tau) - \mathbf{y}_0(\tau))d\tau$$

where

$$\mathbf{y}_{1/2}(t) = \mathbf{y}_0 + \int_0^t \mathbf{f}(\mathbf{y}_0(\tau))d\tau,$$

EXAMPLE We apply this to the example above with $y_0(t) \equiv 1$. We find

$$y_1(t) = 1 + t + t^2,$$
$$y_2(t) = 1 + t + t^2 + t^3 + t^4 + \tfrac{4}{5}t^5 + \dots,$$

less accurate than the pure Newton iteration but easier to compute. Figure

5–1 compares the Picard, modified Newton, and pure Newton iterations on this problem. ●

	Picard	Modified Newton	Pure Newton
$y_0(t)$	1	1	1
$y_1(t)$	$1+t$	$1+t+t^2$	$1+t+t^2+\frac{2}{3}t^3+\ldots$
$y_2(t)$	$1+t+t^2+\frac{1}{3}t^3$	$1+t+t^2+t^3+t^4+\frac{4}{5}t^5+\ldots$	$1+t+t^2+t^3+t^4+t^5+$ $+t^6+\frac{62}{63}t^7+\ldots$

FIGURE 5–1

In practice, of course, one would have to solve the linear differential or integral equations in the pure method or evaluate the integrals in the modified method numerically; the total error will then depend both on the stage of the iteration and the accuracy with which it is performed, but convergence still can be proved.

Additional references: 3, 44.

5.8 ERROR-BOUNDING METHODS

As we have mentioned before, several methods allow for *estimation* and *control* of the local error. For example, a further evaluation of the right-hand side can be used for Runge-Kutta methods to estimate error, and for predictor-corrector methods the difference between the predicted and corrected values can be used. These methods do not, however, *bound* the local error, let alone the global error; it is sometimes desirable to have such bounds. Some work has been done in this direction using majorization ideas [33], but this has not yet been thoroughly developed for computational purposes. One of the oldest error-bounding methods for equations for y in E^1 is based on a result of Chaplygin [19]. If $u(t_0) = v(t_0) = y(t_0)$ and

$$u' - f(t, u) < 0 \qquad \text{for } t_0 < t < t_1,$$

$$v' - f(t, v) > 0 \qquad \text{for } t_0 < t < t_1,$$

$$y' - f(t, y) = 0 \qquad \text{for } t_0 < t < t_1,$$

then

$$u(t) < y(t) < v(t) \qquad \text{for } t_0 < t < t_1.$$

If $(\partial^2 f)/(\partial y^2)$ is of one sign for $u(t) \leq y(t) \leq v(t)$ and $t_0 \leq t \leq t_1$, then [80] the solutions u_1, v_1 of

$$u_1' = f(t, u) + \frac{f(t, v) - f(t, u)}{v - u}[u_1 - u], \qquad u_1(t_0) = y(t_0),$$

$$v_1' = f(t, u) + \frac{\partial f}{\partial y}(t, u)[v_1 - u], \qquad v_1(t_0) = y(t_0)$$

satisfy

$$u(t) < u_1(t) < y(t) < v_1(t) < v(t), \qquad t_0 < t < t_1;$$

this iteration bounding y can be continued.

EXAMPLE Consider the problem $y' = y^2$, $y(0) = 1$. Consider trial functions of the form $z(t) = 1 + at$. Then $z' - z^2 = a - (1 + 2at + t^2)$. If $a = 1$, then $z' - z^2 < 0$ for $t > 0$. If $a > 1$, $z' - z^2 > 0$ for $0 < t < (-1 + \sqrt{a})/a$. The interval is of maximum length with $a = \sqrt{2}$, in which case we find

$$1 + t < y(t) < 1 + \sqrt{2}\,t \qquad \text{for } 0 < t < \frac{-1 + 2^{1/4}}{2^{1/2}} \leq 0.142.$$

The solution is of course $y(t) = 1/(1 - t) = 1 + t + t^2 + \dots$. ●

One of the more recent methods of error bounding is that of ellipsoidal error bounds [62, 63]. With this method the true solution $\mathbf{y}(t)$ is guaranteed to lie inside an ellipse centered about the computed solution $\mathbf{Y}(t)$; the quadratic form of the ellipse is described by a matrix $\mathbf{M}(t)$ which is itself computed from a system of roughly $n^2/2$ differential equations, where $\mathbf{y}$ lies in E^n. The nature of the computations involving $\mathbf{M}(t)$ are straightforward enough that, surprisingly, solving the large system determining $\mathbf{M}(t)$ does not appear to add unreasonably to the time necessary to compute $\mathbf{y}(t)$ alone for many problems of interest. Thorough analysis of the application of the method to nonlinear systems is not available at this writing, but preliminary evidence indicates that it will be of great value when it is better understood and its application has been perfected.

A somewhat similar method, some of whose disadvantages prompted the development of the ellipsoidal bounding technique, uses *interval analysis* to generate bounds in the form of rectangular parallelepipeds having faces parallel to the coordinate planes in E^n. We examine this automated and nearly perfected method further.

First we must understand the basic ideas of interval arithmetic. In trying to find error bounds for an answer, we really seek an *interval* in which that answer lies; to use this answer in further computations, we need to be able to perform arithmetic on intervals. Therefore we consider interval numbers (vectors) of the form

$$\mathbf{I} = [\mathbf{a}, \mathbf{b}] = ([a_1, b_1], \ldots, [a_n b_n]).$$

The result of any arithmetic operation $*$ on two intervals $\mathbf{I}_1$ and $\mathbf{I}_2$ should be defined as

$$\mathbf{I}_1 * \mathbf{I}_2 = \{\mathbf{x}; \mathbf{x} = \mathbf{x}_1 * \mathbf{x}_2, \mathbf{x}_1 \in \mathbf{I}_1, \mathbf{x}_2 \in \mathbf{I}_2\}.$$

On a computer, however, this interval might not be exactly representable. For that reason we must use *rounded* interval arithmetic in which to the operation $\mathbf{I}_1 * \mathbf{I}_2$ the computer assigns some machine-representable interval $\mathbf{I}$ containing the exact interval result. More generally, for any function $\mathbf{f}(\mathbf{x})$ we call $\mathbf{F}(\mathbf{X})$ an *interval extension* if, for any interval $\mathbf{X}$, $\mathbf{F}(\mathbf{X})$ is a machine-representable interval containing $\{\mathbf{y}; \mathbf{y} = \mathbf{f}(\mathbf{x}), \mathbf{x} \in \mathbf{X}\}$.

Now consider the differential equation

$$\mathbf{y}' = \mathbf{f}(\mathbf{y}), \qquad \mathbf{y} \in E^n, \qquad \mathbf{y}(0) = \mathbf{y}_0. \tag{5–9}$$

We shall compute guaranteed bounds for the solution by an interval version of Taylor's formula. Suppose that we have n-dimensional intervals $\mathbf{Y}_0$ and $\mathbf{A}$, that is, n-dimensional rectangles, such that $\mathbf{y}_0 \in \mathbf{Y}_0 \subset \mathbf{A}$ and $\mathbf{y}_0$ is in the interior of $\mathbf{A}$. Suppose that we have interval extensions $\mathbf{F}^{(j)}(\mathbf{Y})$ of the total derivatives $\mathbf{f}^j(\mathbf{y}) = (d^{j+1}/dt^{j+1})\mathbf{y}(t)$ of solutions to Eq. 5–9, that is

$$\mathbf{f}^0(\mathbf{y}) = \mathbf{f}(\mathbf{y}),$$

$$\mathbf{f}^1(\mathbf{y}) = \frac{\partial \mathbf{f}}{\partial \mathbf{y}} \cdot \mathbf{f}(\mathbf{y}),$$

et cetera. If we expand $\mathbf{y}(t)$ in a Taylor series with remainder about $t = 0$ and replace derivatives by interval extensions, we can consider the interval function

$$\mathbf{Y}(t) = \mathbf{Y}_0 + \sum_{j=1}^{K-1} \frac{\mathbf{F}^{j-1}(\mathbf{Y}_0)}{j!}\, t^j + \frac{F^{K-1}(\mathbf{A})}{K!}\, t^K.$$

It can be shown [81] that, under reasonable assumptions on the $\mathbf{f}^j$ and $\mathbf{F}^j$, if $\mathbf{Y}(t) \subset \mathbf{A}$ for $0 \leq t \leq h$, then the solution $\mathbf{y}(t)$ satisfies $\mathbf{y}(t) \in \mathbf{Y}(t)$ for

$0 \leq t \leq h$. It is possible to program a computer to generate the formal derivatives $\mathbf{f}^j$ and their interval extensions and also to find h and $\mathbf{A}$ such that $\mathbf{Y}(t) \subset \mathbf{A}$ for $0 \leq t \leq h$. This then yields bounds on the solution for $0 \leq t \leq h$ and we start the process again now at $t = h$, with $\mathbf{Y}_0$ replaced by $\mathbf{Y}(h)$. The method has been implemented for the CDC 3600 [11].

EXAMPLE Consider again the equation

$$y' = y^2, \qquad y \in E^1, \qquad y(0) = 1$$

having exact solution $y(t) = 1/(1 - t)$. Given ϵ and K, set

$$h = \tfrac{1}{2}(\epsilon/[K + 1])^{1/(K+1)};$$

if ϵ is small enough that $h \leq 1$, then the interval

$$A = [1, 1 + (2h) + (2h)^2 + \ldots + (2h)^{k-1} + (2h)^k(1 + 2h)^{k+1}]$$

is such that $Y([0, h]) \subset A$ and therefore $Y(t)$ bounds the solution. For $\epsilon = 10^{-9}$ and $K = 9$ we have $h = 0.05$, which yields bounds on the solution for $0 \leq t \leq 0.05$; the width of the bounding interval $Y(t)$ is bounded by 10^{-11}. A SHARE program (ML MDJS, SDA #3210), written to implement the methods in general on an IBM 7094, was applied to this problem; selected data are given in Figure 5–2.

t	Interval midpoint	Error bound
0.044199569	1.0462435	$7.4505806 \cdot 10^{-8}$
0.17553912	1.2129139	$1.1920929 \cdot 10^{-7}$
0.92183263	12.793067	$1.5974045 \cdot 10^{-5}$
0.99986639	7486.0633	$5.6754150 \cdot 10^{0}$

FIGURE 5–2

We remind the reader that the true solution blows up at $t = 1$; the true solution at $t = 0.99986639$ is 7484.5, correctly rounded. ●

If one wants guaranteed error bounds, the above method can provide them for many problems.

5.9 ON CHOOSING A METHOD FOR INITIAL VALUE PROBLEMS

In the first half of Chapter 4 and throughout this chapter we have given brief descriptions of many of the types of methods used at present or recently developed and proposed for use to solve initial value problems; the reader might well ask: "Which one should I use?" The answer commonly given (and one which we are inclined to give) is: "It all depends; consult an expert!" Since experts often are not immediately available to the person needing an answer, and since even they might have a difficult time, we propose now to state some of our personal *opinions* and *judgments* about what to do in general; this section might therefore be entitled: "What to Do Until the Numerical Analyst Comes." We recognize that our opinions will be neither all correct nor all universally accepted by our colleagues; despite this, we feel it is desirable at least to attempt some discussion of possible guidelines for choosing a method.

Historically, most people seem to have chosen methods of order roughly equal to four, independent of the type of method used; the most popular and widely used Runge-Kutta and predictor-corrector methods are of about this order. This phenomenon appears to be a remnant of the days of hand computation and desk computers, when it was observed that divided differences of order greater than about four using eight- or ten-digit arithmetic often contained much "noise" and few significant figures; also, few higher-order one-step methods were generally known, and for some of those that were, the values of the coefficients were not entirely to be trusted. We tend to believe that even today it is better in general not to use finite-difference multistep methods of order appreciably greater than four when only eight- or ten-decimal digits are carried; some extensive experiments have indicated [58] that fifth-order methods are often best, supporting our contention. We qualify this, however, with the remark that the new multistep methods mentioned in Section 5.5 which achieve high-order accuracy with the same number of points, one off the grid, may well raise this number to roughly seven for computing very smooth solutions when more experimental evidence is known. Despite the fact that techniques are known for varying the step size in multistep methods, our personal experience indicates that such methods are not really appropriate for problems in which for efficiency the step size may need to vary often or greatly or at unknown places, as in some space-flight problems; for such problems we prefer a variable-step-size implementation of one-step methods. However, if step change is not needed for efficiency, we prefer predictor-corrector methods in general. A reasonable compromise is to use different methods for different ranges of the solution, depending on the behavior of the solution.

If the system of equations is such that the computer can derive subroutines equivalent to recursive formal differentiations to yield values for higher derivatives of the unknowns, recent investigations seem to indicate that methods using such information are to be preferred, at least for computing very smooth solutions: Lie series, Taylor series, and the Fehlberg modification of Runge-Kutta which we will discuss in Section 9.3 are methods of this type. For significant problems, comparative studies have shown that accuracy, the number of right-hand-side evaluations, and computer time, including time for formal differentiation, are optimized by using such methods—usually of rather high order, such as roughly nine for ten-digit arithmetic or about sixteen for eighteen digits. For some particular solutions of the restricted three-body problem, for example, the Lie-series and Fehlberg methods proved superior to standard multistep and Runge-Kutta methods [67]; a special multistep method using off-grid points was also shown to be excellent for this problem [39, 46].

When choosing a method for a given problem, it must be decided what stability properties are needed. If a solution to a stiff equation is needed over very large time, it probably is wise to use one of the super-stable methods with properties discussed in Section 5.4; it is not yet possible to make comparisons among the few particular methods available. For more tractible problems an S-stable method is almost always needed, not just a D-stable one; we recommend this without heistation. For the occasional problem with severe inherent stability difficulties but for which asymptotic behavior of the particular solution is known, we highly recommend the boundary-value approach of Section 5.3.

We do not recommend that integral-equation methods be used on a computer when the methods already mentioned are available. They do appear to be of some value, however, for crude approximations in general or for accurate approximations for small time intervals, and may thus be of value for providing first approximations for a more powerful approach, such as that given by Lie series. We feel much the same way about methods of best fit; however, these do appear to be of use for *boundary* value problems and were presented in this chapter chiefly for that reason, but we do not recommend them for initial value problems.

Acceleration methods such as those discussed in Section 4.12 can be very valuable when applicable; however, it may sometimes be too expensive—that is, require too much computer time—to solve a given problem by using different step sizes in order to extrapolate to the limit. Deferred corrections can be used to increase accuracy also, especially when formal differentiations have provided the high derivatives necessary for use in this approach.

Generally it is not necessary to have guaranteed error bounds for solutions; however, when it is necessary, the automated methods of interval analysis, based on high-order Taylor-series expansions, are preferable to those using ideas of majorization or monotonicity as involved in Chaplygin's method.

Recently, several authors have attempted to define what is meant by a best method and to formalize the comparative analysis of methods as applied to specific classes of problems [45, 57, 59, 60, 96]. We feel that this is a fruitful field for research but that as yet no great progress has been achieved with respect to suggestions concerning the method to use. Most of the numerical methods we have mentioned are *reasonably* efficient, even if not best in every sense; each has its own advantages and disadvantages, although some of the former and, fortunately, most of the latter, are *minor*. One of our contentions in Part III, however, is that, no matter what method of computation is chosen, there are very often *major* advantages to be gained from transforming a given problem by using knowledge, however obtained, about the behavior of the desired solution.

Of course, even for a *single given* differential equation it is not usually possible to specify a best single method to use or a good choice of transformed variables independent of the initial or boundary conditions and length of time over which the solution is to be computed. For example, there are reputed to be well over a thousand published papers on the behavior of solutions for the restricted three-body problem alone, and on methods for computing and studying such solutions; solutions to this fourth-order system constitute a fantastic variety of curves in E^4, the phase space [97]. Thus a good choice of a method for computing or studying a particular solution may depend more on the nature of its detailed behavior and the vector field near it than on the form of the differential equation; it is this, of course, that makes any knowledge at all about the nature of a solution extremely vital to the intelligent choice of a method.

6

SPECIAL NUMERICAL METHODS FOR BOUNDARY VALUE PROBLEMS

6.1 INTRODUCTION

In this chapter we present a brief sketch of a number of interesting types of computational methods for boundary value problems, just as we did for initial value problems in the preceding chapter. Most of these methods have been developed only recently or have only recently been applied to boundary value problems. In some cases theoretical approaches of earlier chapters have been implemented numerically, resulting in the techniques presented here; indicating this strong interplay between theoretical aspects of the equations and valuable numerical methods of solution is, of course, one of the aims of this book.

6.2 METHODS OF BEST FIT

The methods described in Section 5.2 for initial value problems can be used without essential change to solve the boundary value problem

$$\mathbf{y}' = \mathbf{f}(t, \mathbf{y}), \qquad \mathbf{y} \in E^n, \qquad 0 \leq t \leq T$$

with boundary conditions

$$\mathbf{My}(0) + \mathbf{Ny}(T) = \mathbf{k} \in E^n.$$

Thus, given J functions $\varphi_j(t) \in E^n$, we seek $\boldsymbol{\alpha} \in E^J$ such that

$$\boldsymbol{\phi}(t; \boldsymbol{\alpha}) = \sum_{j=1}^{J} \alpha_j \varphi_j(t)$$

minimizes $\|\boldsymbol{\phi}'(t; \boldsymbol{\alpha}) - \mathbf{f}(t, \boldsymbol{\phi}(t; \boldsymbol{\alpha}))\|_L$ over the set of $\boldsymbol{\alpha}$ satisfying

$$\mathbf{M}\boldsymbol{\phi}(0; \boldsymbol{\alpha}) + \mathbf{N}\boldsymbol{\phi}(T; \boldsymbol{\alpha}) = \mathbf{k},$$

where $\|\mathbf{w}(t)\|_L$ is some nonnegative function (semi-norm) based on the values $\|\mathbf{w}(t_i)\|$, $0 = t_0 < t_1 < \ldots < t_L = T$.

In some other cases, generally for linear homogeneous equations, it may be that $\boldsymbol{\phi}(t; \boldsymbol{\alpha})$ solves the differential equation exactly for certain values of $\boldsymbol{\alpha}$ and that the problem is so to choose $\boldsymbol{\alpha}$ from among those values as to satisfy the boundary conditions (perhaps nonlinear) accurately. For clarity of exposition we consider the special problem

$$y'' = f(t, y, y'), \qquad y \in E^1, \qquad 0 \le t \le T$$

with boundary conditions

$$y(0) = y(T) = 0.$$

If $J = L + 3$, then often $\boldsymbol{\alpha}$ can be found such that

$$\|\phi''(t; \boldsymbol{\alpha}) - f(t, \phi, \phi')\|_L = 0,$$

$$\phi(0; \boldsymbol{\alpha}) = \phi(T; \boldsymbol{\alpha}) = \mathbf{0},$$

in which case we have a *collocation* method. The situations with $J < L + 3$ or $J > L + 3$ yield mathematical programming problems or problems of choosing smooth approximations, just as for initial value problems; the approximations can be shown to converge under suitable hypotheses.

Additional references: 36, 68, 89, 90.

6.3 VARIATIONAL METHODS

As we observed in Section 3.6, many boundary value problems of the form

$$y'' = f(t, y, y'), \qquad y \in E^1, \qquad 0 \le t \le T$$

$$y(0) = \alpha, \qquad y(T) = \beta$$

are equivalent to variational problems of the form:

$$\text{minimize} \qquad J(y) = \int_0^T F(t, y(t), y'(t))dt$$

over a set of functions y having one continuous derivative and satisfying $y(0) = \alpha$, $y(T) = \beta$. Although it is true more generally that higher-order equations (that is, involving higher derivatives of y) can be similarly treated, for simplicity we restrict ourselves to the second-order case to present briefly a numerical method based on the variational formulation of the problem. The method is essentially similar to those of the preceding section; the error measure now, however, is more closely related to the equation itself—that is, we measure the error by the deviation of $J(y)$ from its minimum and try to force the deviation to zero. For simplicity, we suppose that the boundary values α and β satisfy

$$\alpha = \beta = 0.$$

The basic method is that of *Ritz*. Consider K functions $\varphi_j(t)$ defined on $[0, T]$ and satisfying

$$\varphi_j(0) = \varphi_j(T) = 0, \qquad j = 1, 2, \ldots, K.$$

We seek to find $\boldsymbol{\alpha} \in E^K$ so that

$$\phi(t; \boldsymbol{\alpha}) \equiv \sum_{j=1}^K \alpha_j \varphi_j(t)$$

minimizes $J(\phi(.; \boldsymbol{\alpha}))$ over all $\boldsymbol{\alpha}$. If the solution $y(t)$ to our problem can be accurately approximated by linear combinations of the functions φ_j, we would expect the minimizing element $\phi(t; \boldsymbol{\alpha}_K)$ to converge to y as K tends to infinity.

This basic method and similar or equivalent forms such as the Galerkin methods, the orthogonal-projection method, the Trefftz method, and others have been known for many years but are now being revived for computation. From a naive viewpoint we might say that the basic physical problem gave rise to the variational problem, and it is that which we should deal with numerically, rather than the differential equation created by man to simplify his analysis; from this viewpoint we should approximate the energy of the system, not its derivative. To demonstrate the method, we consider some recent results and their application to a particular problem.

Consider the equation

$$y'' = f(t, y), \qquad 0 \leq t \leq 1$$

with boundary conditions

$$y(0) = y(1) = 0.$$

Suppose $(\partial^2 f/\partial y^2) \geq 0$ for all y and $0 \leq t \leq 1$. This problem is equivalent to the variational problem with

$$F(t, y, y') = \frac{-y'^2}{2} + \int_0^y f(t, \eta)d\eta.$$

It can be shown [21] that if the solution $y(t)$ is at least twice continuously differentiable on [0, 1], that is, $y \in C^p, p \geq 2$, if there exist functions $\phi(t; \boldsymbol{\alpha}_K')$ for some $\boldsymbol{\alpha}_K' \in E^K$ such that

$$\lim_{K\to\infty} \int_0^1 (y'(t) - \phi'(t; \boldsymbol{\alpha}_K'))^2 dt = 0,$$

then the function $\phi(t; \hat{\boldsymbol{\alpha}}_K)$ minimizing $J(\phi(t; \boldsymbol{\alpha}))$ for $\boldsymbol{\alpha} \in E^K$ converges uniformly to y and satisfies

$$\max_{0\leq t\leq 1} |y(t) - \phi(t; \hat{\boldsymbol{\alpha}}_K)| \leq C \inf_{\alpha\in E^K} [\int_0^1 (y'(t) - \phi'(t; \boldsymbol{\alpha}))^2 dt]^{1/2}$$

for some constant C. In particular, if

$$\varphi_j(t) = t^{j-1}, \qquad j = 1, 2, \ldots, K,$$

then there is a constant M such that

$$\max_{0\leq t\leq 1} |y(t) - \phi(t; \hat{\boldsymbol{\alpha}}_K)| \leq M \frac{1}{(K-2)^{p-1}};$$

that is, the minimizing polynomial of degree n converges to the solution uniformly like $(1/(n-1))^{p-1}$. Similar results for particular sets of approximating functions $\varphi_j(t)$ can also be deduced, so long as it is possible to approximate y' in a square-integral sense. If the solution is very smooth, then theoretical and experimental results indicate [21] that very accurate solutions can be obtained with polynomial approximation. For solutions with certain discontinuities in derivatives, other types of approximating functions, such as splines, will be more valuable. Part of the power in this variational approach lies in the ability to use approximate solutions which can exhibit the same kind of characteristics that the exact solution can.

EXAMPLE [21] Consider the problem

$$y'' = e^y, \qquad y(0) = y(1) = 0;$$

we seek to minimize $\int_0^1[-(y'^2/2)+e^y]dt$ over the set of polynomials of degree n vanishing at 0 and 1. Replacing the integral by an appropriate quadrature scheme, we find that the computed polynomial approximates the solutions to within 5×10^{-4} for $n=3$, 4×10^{-6} for $n=5$, and 5×10^{-8} for $n=7$. ●

As in the above example, one must always replace the integral by some finite sum; if the accuracy of the quadrature formula is high enough, the overall accuracy is not decreased and convergence to the solution still takes place.

Additional references: 9, 22, 23, 40, 104.

6.4 LIE-SERIES METHODS

In Section 3.3 we observed that the formal Lie-series method of Section 2.4 could be used to reduce a boundary value problem to a system of equations for an unknown initial vector $\mathbf{y}_0$. Just as Lie series are of value for solving initial value problems, they may well become a tool for boundary value problems. We illustrate by an example.

EXAMPLE Consider the problem

$$y''=e^y, \qquad y(0)=y(1)=0, \qquad y\in E^1.$$

We convert this to the system

$$y_1'=y_2$$

$$y_2'=e^{y_1}$$

and use Lie series, based on an unknown initial vector $\mathbf{c}$. The operator $\mathbf{F}$ is given by

$$F=y_2\frac{\partial}{\partial y_1}+e^{y_1}\frac{\partial}{\partial y_2},$$

and, setting $\mathbf{Y}=\begin{pmatrix}y_1\\y_2\end{pmatrix}$, we have

$$\mathbf{F}(\mathbf{Y})=\begin{pmatrix}y_2\\e^{y_1}\end{pmatrix},$$

$$\mathbf{F}^2(\mathbf{Y})=\begin{pmatrix}e^{y_1}\\y_2e^{y_1}\end{pmatrix},$$

$$\mathbf{F}^3(\mathbf{Y}) = \begin{pmatrix} y_2 e^{y_1} \\ y_2^2 e^{y_1} + e^{2y_1} \end{pmatrix},$$

et cetera. According to the Lie-series formula,

$$\mathbf{Y}(t) = [e^{t\mathbf{F}}\mathbf{Y}]_{\mathbf{Y}=\mathbf{c}},$$

and we want the first component y_1 of $\mathbf{Y}$ to vanish at $t = 0$ and $t = 1$. If we truncate the series for $e^{t\mathbf{F}}$ after N terms and set the first component equal to zero to solve for $y_2(0) = c_2$, we find the following approximate solutions.

$N = 1$. We take $\mathbf{Y}(t) = \begin{pmatrix} 0 \\ c_2 \end{pmatrix}$, so $c_2 = 0$ and $y(t) \sim 0$.

$N = 2$. We take $\mathbf{Y}(t) = \begin{pmatrix} 0 \\ c_2 \end{pmatrix} + t\begin{pmatrix} c_2 \\ 1 \end{pmatrix}$, again yielding $c_2 = 0$ and $y(t) \sim 0$.

$N = 3$. We take $\mathbf{Y}(t) = \begin{pmatrix} 0 \\ c_2 \end{pmatrix} + t\begin{pmatrix} c_2 \\ 1 \end{pmatrix} + \frac{t^2}{2}\begin{pmatrix} 1 \\ c_2 \end{pmatrix}$, which yields $c_2 = -\frac{1}{2}$ and $y(t) \sim (t^2 - t)/2$.

$N = 4$. We take $\mathbf{Y}(t) = \begin{pmatrix} 0 \\ c_2 \end{pmatrix} + t\begin{pmatrix} c_2 \\ 1 \end{pmatrix} + \frac{t^2}{2}\begin{pmatrix} 1 \\ c_2 \end{pmatrix} + \frac{t^3}{6}\begin{pmatrix} c_2 \\ c_2^2 + 1 \end{pmatrix}$, which yields $c_2 = -\frac{3}{7}$, $y(t) \sim -\frac{3}{7}t + \frac{t^2}{2} - \frac{t^3}{14}$.

Thus, truncating the Lie series yields a computable approximate solution with a reasonable accuracy. The maximum error in the above example, for instance, is less than 0.015 for the low value of $N = 3$; for $N = 4$, however, the error is at least 0.016, indicating that the error need not decrease monotonically as N increases. For $N = 5$ the error decreases to 0.004. ●

6.5 MONOTONICITY METHODS

In Section 3.7 we observed that certain differential operators of the form

$$\mathbf{D}(\mathbf{u}) \equiv -\mathbf{u}'' + \mathbf{f}(t, \mathbf{u}, \mathbf{u}')$$

with general linear boundary conditions given by an operator $\mathbf{B}(\mathbf{u})$ were of monotone kind, that is, from $\mathbf{D}(\mathbf{u}) \geq \mathbf{D}(\mathbf{v})$ and $\mathbf{B}(\mathbf{u}) \geq \mathbf{B}(\mathbf{v})$ it follows that $\mathbf{u} \geq \mathbf{v}$. Thus, if $\mathbf{y}$ solves $\mathbf{D}(\mathbf{y}) = 0$, $\mathbf{B}(\mathbf{y}) = 0$, then $\mathbf{D}(\mathbf{u}) \leq 0 \leq \mathbf{D}(\mathbf{v})$ and

$\mathbf{B}(\mathbf{u}) \leq 0 \leq \mathbf{B}(\mathbf{v})$ imply $\mathbf{u} \leq \mathbf{y} \leq \mathbf{v}$. This can be used to generate approximate solutions.

EXAMPLE We saw in Section 3.7 that the problem

$$y'' = e^y, \qquad y(0) = y(1) = 0, \qquad y \in E^1$$

could be treated in this manner; we now compute functions u and v bounding y. Consider $w(t) = at^2 + bt + c$; to satisfy the boundary conditions we need $c = w(0) = 0$, $a + b = w(1) = 0$. Thus we consider $w(t) = a(t^2 - t)$. We have $-w'' + e^w = -2a + e^{a(t^2-t)}$. Since $t^2 - t \leq 0$ on $[0, 1]$ and the solution $y(t)$ was found in Section 3.7 to satisfy $y(t) \leq 0$, we want to choose $a > 0$. Then the minimum value of $e^{a(t^2-t)}$ occurs at $t = \frac{1}{2}$ and equals $e^{-a/4}$. To make $-w''+e^w \geq 0$, we need $-w''+e^w = -2a+e^{a(t^2-t)} \geq -2a+e^{-a/4} \geq 0$; looking in a table of exponentials we see that $-2(0.44) + e^{-.11} \geq -0.88 + 0.89 = 0.01 \geq 0$. Therefore the solution $y(t)$ satisfies $y(t) \leq 0.44(t^2 - t)$. Similarly, choosing a so that $-w'' + e^w = -2a + e^{a(t^2-t)} \leq -2a + e^0 = -2a + 1 \leq 0$, we take $a = 0.5$ which gives $-w'' + e^w \leq 0$. In summary,

$$-0.5(t - t^2) \leq y(t) \leq -0.44(t - t^2), \qquad 0 \leq t \leq 1;$$

$y(t)$ has been trapped in an interval of width at most 0.0015. ●

It is also possible to give iterative methods that generate sequences $\mathbf{u}_n(t)$ and $\mathbf{v}_n(t)$ satisfying

$$\mathbf{u}_0 \leq \mathbf{u}_1 \leq \ldots \leq \mathbf{y} \leq \ldots \leq \mathbf{v}_1 \leq \mathbf{v}_0.$$

For example, consider the problem

$$y'' = f(t, y), \qquad y(0) = y(1) = 0, \qquad y \in E^1,$$

where $(\partial f/\partial y) \geq 0$. We might consider attempting to solve this by choosing a function $y_0(t)$ and then iteratively defining $y''_{n+1} = f(t, y_n)$, $y_{n+1}(0) = y_{n+1}(1) = 0$; this essentially is Picard's method and in the present situation can generate monotone sequences. Specifically suppose we have $u_0(t)$ satisfying

$$-u_0'' + f(t, u_0) \leq 0, \qquad u_0(0) = u_0(1) = 0.$$

For $n = 0, 1, \ldots$ find functions $v_n(t)$ and $u_{n+1}(t)$ satisfying

$$v_n'' = f(t, u_n), \qquad v_n(0) = v_n(1) = 0,$$

$$u_{n+1}'' = f(t, v_n), \qquad u_{n+1}(0) = u_{n+1}(1) = 0.$$

It is easy to show that $u_n \leq y \leq v_n$ for all n. More importantly, if $u_0 \leq u_1$, then we have

$$u_0 \leq u_1 \leq u_2 \leq \ldots \leq y \leq \ldots \leq v_2 \leq v_1 \leq v_0.$$

EXERCISE

Apply the technique just described to the example $y'' = y^3 - 1$, $y(0) = y(1) = 0$. Using $u_0(t) \equiv 0$, obtain $v_0(t)$ and $u_1(t)$ explicitly and plot them together.

It is possible to generate such bounding sequences in a different fashion by using Newton's method, if $\partial^2 f/\partial y^2$ is of constant sign; in this case Newton's method is often called *quasi-linearization.*

For monotone methods based either on Picard or Newton iterations, one has to solve a linear problem at each step. Under certain hypotheses it can be shown that the numerical solutions also bound $y(t)$ in the same fashion.

Additional references: 8, 19, 25, 26, 27, 93, 98.

6.6 ERROR-BOUNDING METHODS

The methods of the preceding section provide a way to compute guaranteed bounds on some solutions; such bounds can be made to include computational error, that is rounding error, by using interval arithmetic. For cases in which monotone methods are inapplicable, it is still possible to compute guaranteed bounds via interval methods. Let us discuss briefly two possible approaches.

Consider equations of the form

$$y'' = f(t, y, y'), \qquad 0 \leq t \leq 1, \qquad y \in E^1 \tag{6–1}$$

with $y(0)$ and $y(1)$ given. If we discretize the problem with a step size of $h = 1/N$, so that $0 = t_0 < t_1 < \ldots < t_N = 1$, $t_{i+1} - t_i = h$, then we know that

$$y''(t_i) = \frac{y(t_{i+1}) - 2y(t_i) + y(t_{i-1})}{h^2} - \frac{h^2}{12}\, y^{(4)}(\theta_{1i}),$$

$$y'(t_i) = \frac{y(t_{i+1}) - y(t_{i-1})}{2h} - \frac{h^2}{6}\, y^{(3)}(\theta_{2i}), \tag{6–2}$$

$$t_{i-1} < \theta_{1i},\ \theta_{2i} < t_{i+1}.$$

Functions $p(t, y, y')$ and $q(t, y, y')$ can be computed from the differential equation such that

$$y^{(4)}(t) = p(t, y, y'),$$

$$y^{(3)}(t) = q(t, y, y').$$

Thus, if one has crude bounds on $y(t)$ and $y'(t)$ and interval extensions of p and q can be found, then refined bounds for $y(t)$ and $y'(t)$ can be computed by carrying the interval bounds along in the finite-difference method based on Eq. 6–2 with $f(t_i, y(t_i), y'(t_i))$ substituted for y''. It has been shown [54] that such preliminary bounds can be computed; these bounds are quite sharp if the equation is linear but may be rather crude otherwise. If one must have guaranteed bounds, this approach appears to be successful.

There is another interval method, however, which often requires less computation. Consider Eq. 6–1 again; suppose that one has obtained an approximate solution by some means and now wishes to compute error bounds. Given a reasonably accurate first approximation, perhaps the best way to get an improved approximation is by Newton's linearization method with its error-squaring behavior. For our purposes, Newton's method has a further advantage; there exist known theorems of so-called Kantorovich type which guarantee that, under certain hypotheses, a solution exists within a specific computable distance from the initial approximation. It has recently been shown [99] that by interval analysis it is possible to discover cases in which the needed hypotheses hold and to compute the maximum distance to a solution. Thus, simultaneously, one computes a new approximation by Newton's method and generates guaranteed error bounds. A computer program implementing this has been written and tested, and has given accurate results.

6.7 FIRST APPROXIMATIONS

In many of the numerical methods for boundary value problems, a reasonable first approximation to the solution makes subsequent computations much easier and requires fewer such computations; this is particularly true for such iterative procedures as Newton's for shooting or integral-equation methods and for Lie-series and error-bounding techniques. A reasonable first approximation is also useful to start an iterative process for solving the finite-difference equations when they are used. Here "reasonable" is taken to mean, "At the worst, having roughly the same shape as the true solution."

Often a reasonable approximation can be obtained from best-fit or variational methods with few parameters or from monotonicity techniques. Another approach is to use finite-difference methods with a sequence of step sizes which get smaller, using the preceding solution as initial guess each time for the next step size.

More generally, one might use *homotopy* or *successive-perturbation* methods in which the equation

$$\mathbf{y}'' = \mathbf{f}(t, \mathbf{y}; \mathbf{y}')$$

is replaced by

$$\mathbf{y}'' = \mathbf{f}(t, \mathbf{y}, \mathbf{y}'; \lambda)$$

in which the parameter λ is introduced artificially so that when $\lambda = 0$ we can solve very easily, and when $\lambda = 1$ we have precisely our original equation; we then solve the sequence of problems approximately

$$\mathbf{y}'' = \mathbf{f}(t, \mathbf{y}_i, \mathbf{y}_i'; \lambda_i)$$

for $0 = \lambda_0 < \lambda_1 < \ldots < \lambda_N = 1$ using the final approximation to the solution $\mathbf{y}_i$ to supply an initial approximation to $\mathbf{y}_{i+1}$. One can often find a differential equation for $\mathbf{y}$ as a function of λ and use initial value methods to solve that equation.

EXAMPLE Consider the system

$$\mathbf{y}'' = \mathbf{A}\mathbf{y} + \mathbf{g}(\mathbf{y}, \mathbf{y}'), \qquad \mathbf{y} \in E^n, \qquad \mathbf{y}(0) = \mathbf{c}, \qquad \mathbf{y}(1) = \mathbf{d};$$

we introduce the artificial parameter λ via

$$\mathbf{y}'' = \mathbf{A}\mathbf{y} + \lambda\mathbf{g}(\mathbf{y}, \mathbf{y}'), \qquad \mathbf{y}(0) = \mathbf{c}, \qquad \mathbf{y}(1) = \mathbf{d}.$$

Choose N large and let $\lambda_i = i/N$, $i = 0, 1, \ldots, N$, $\epsilon = 1/N$. If we suppose that it is easy to compute the solution to the system

$$\mathbf{y}_0'' = \mathbf{A}\mathbf{y}_0 = \mathbf{A}\mathbf{y}_0 + \lambda_0\mathbf{g}(\mathbf{y}_0, \mathbf{y}_0'), \qquad \mathbf{y}_0(0) = \mathbf{c}, \qquad \mathbf{y}_0(1) = \mathbf{d},$$

then the solution $\mathbf{y}_1(t)$ to

$$\mathbf{y}_1'' = \mathbf{A}\mathbf{y}_1 + \lambda_1\mathbf{g}(\mathbf{y}_1, \mathbf{y}_1') = \mathbf{A}\mathbf{y}_1 + \epsilon\mathbf{g}(\mathbf{y}_1, \mathbf{y}_1'), \qquad \mathbf{y}_1(0) = \mathbf{c}, \qquad \mathbf{y}_1(1) = \mathbf{d}$$

can be computed using a perturbation expansion in ϵ; that is, we can let

$$\mathbf{y}_1 = \mathbf{u}_0 + \epsilon\mathbf{u}_1 + \epsilon^2\mathbf{u}_2 + \ldots$$

and find that

$$\mathbf{u}_0(t) = \mathbf{y}_0(t),$$

$$\mathbf{u}_1'' = \mathbf{A}\mathbf{u}_1 + \mathbf{g}(\mathbf{u}_0, \mathbf{u}_0'), \qquad \mathbf{u}_1(0) = \mathbf{0}, \qquad \mathbf{u}_1(1) = \mathbf{0},$$

et cetera. The perturbation solution can be further refined if desired. More generally we can compute the solution $\mathbf{y}_{i+1}(t)$ of

$$\mathbf{y}_{i+1}'' = \mathbf{A}\mathbf{y}_{i+1} + \lambda_{i+1}\mathbf{g}(\mathbf{y}_{i+1}, y_{i+1}') = [\mathbf{A}\mathbf{y}_{i+1} + \lambda_i\mathbf{g}(\mathbf{y}_{i+1}, \mathbf{y}_{i+1}')] + \epsilon\mathbf{g}(\mathbf{y}_{i+1}, \mathbf{y}_{i+1}'),$$

$$\mathbf{y}_{i+1}(0) = \mathbf{c}, \qquad \mathbf{y}_{i+1}(1) = \mathbf{d},$$

by means of a perturbation expansion in ϵ about the previous solution $\mathbf{y}_i(t)$. The function $\mathbf{y}_N(t)$ will be taken as a numerical solution for the original problem, to provide a first approximation for more refined methods. ●

EXERCISE

Investigate the application of the successive-perturbation method to the example $y'' = y^3 - 1$, $y(0) = y(1) = 0$.

It should be observed that finding a good first approximation is equivalent to transforming the equation by some change of variables so that $y(t) \equiv 0$ becomes a good approximation. Thus the material to be presented in Part III can also be viewed as concerning good first approximations.

6.8 ON CHOOSING A METHOD FOR BOUNDARY VALUE PROBLEMS

As we stated in Section 5.9, the best technique for choosing a method for a given problem may be to consult an expert; otherwise, there are some general lines of consideration one can follow. Historically, shooting methods have not been widely advocated in the literature but have been widely used in practice, even though they are not uniformly successful. The fundamental shooting methods cause some difficulty if the equations for determining the correct initial data are very ill-conditioned; also, if first approximations are inaccurate, solutions to the resulting initial value problems may well become infinite or very distant from the other boundary values. Parallel shooting, however, often avoids these problems quite satisfactorily [64]. Shooting methods should therefore be seriously considered whenever a reasonable first approximation to the whole initial data vector is available.

Finite-difference methods have been the most popular throughout the years; as we remarked in Section 5.9, the best of these in most common circumstances seem to be of about order four, for example, the method

$$Y_{i+1} - 2Y_i + Y_{i-1} = \frac{h^2}{12}(f_{i+1} + 10f_i + f_{i-1})$$

mentioned in Section 4.9. For higher-order methods it appears that the use of deferred or difference corrections is of great importance; finite-difference methods are best used in connection with such an acceleration if one needs the very high-order accuracy obtainable in that way.

Monotonicity methods generally are good for obtaining first approximations and for getting an idea of the behavior of solutions; often the convergence of iterative monotone methods is rather too slow for practice, although quasi-linearization is quite good in those cases in which it is applicable. Results from the theory of operators of monotone kind can often be used to bound the error in a numerical solution; this is an important application but does not use monotonicity properties to compute the solution.

At present the Lie-series techniques for boundary value problems have not been well-enough studied in practice to be judged; because of their performance for initial value problems we tend to feel that they will prove to be of value, given a good first approximation, after they have been studied further.

Variational and best-fit methods, however, already seem to be revealing their power. Variational methods, in particular those of high order, are very useful and in some cases yield computable error bounds; they are not always applicable, however. Experimental evidence has not as yet indicated how much effort is involved in solving the finite-dimensional approximating-variational problem, generating a widely applicable quadrature method, evaluating coefficients in the discrete system, et cetera. It appears that these practical aspects can be satisfactorily treated and that the variational method will be a powerful tool, competitive with other useful techniques. Best-fit methods are, in a sense, generalizations of variational methods to different kinds of functionals that are to be minimized, and should be as efficient as the former methods. Much of the theoretical justification for such techniques remains to be discovered, but this will not prevent their use. Both of these kinds of methods provide very smooth approximations whose error is in many cases well spread out over the interval; this is often desirable.

To some extent this last remark is also true of the integral-equation approach; such a method, however, leads to a nonsparse system of equations to solve and for this reason is rather less desirable than, say, the finite-difference method of order four with its tridiagonal system.

PART 3

TRANSFORMATION OF THE EQUATIONS

An enormous amount of computation has been done on solutions of differential equations, particularly since the appearance of stored-program digital computers. It is interesting (as well as somewhat terrifying) to think that all the arithmetic involved in the significant numerical work of Stromgren [97] on the restricted three-body problem (see p. 127) from about 1900 to 1935, would constitute a short computation for a modern electronic computer. To show this, let us engage in a bit of fanciful exaggeration. Suppose that the calculations in all of the papers on the subject were done by a hundred persons engaged full-time for thirty years to operate desk calculators capable of multiplying a pair of ten-decimal-digit numbers in twenty seconds (including the time for pressing all the buttons). If each person worked a forty-hour week fifty weeks out of the year, this would amount to $40 \times 50 \times 100 \times 30 =$ 6,000,000 man-hours of computing with desk calculators. Assuming that the total time for these computations is proportional to the number of multiplications, we deduce the following comparison. Using a figure of 10 microseconds (10^{-5} seconds) for a multiplication by a digital computer—a conservative figure for some existing computers—we find that the machine would require $(10^{-5}/20) \times 6{,}000{,}000 = 3$ hours to do the work.

There is one very important consideration left out of such deliberations, namely the significance of the results. The analysis, planning, and organization of a thirty-year project would have been, beyond question, incomparably more refined than anything yet put into the preparation of a three-hour machine computation. Advances in programming technique have made it possible for almost anyone to engage a computer in a three-hour project calculating some kind of "numerical solutions" to differential equations; unfortunately, a common practice is to substitute redundant computing for mathematical analysis.

Each computational method for solving problems in differential equations involves various "approximation parameters" such as mesh or step size in discrete-variable methods, or degrees of approximating polynomials. The exact solution to the problem is independent of these things and so, if a succession of computer runs is made with various values of the approximation parameters all leading to essentially the same results, *some* confidence in the results is probably warranted on these grounds alone. There are, to be sure, general-purpose problem-solving computer programs with a fair amount of careful analysis built in. We can take advantage of the tremendous speed and versatility of stored-program computers not only by using redundant computing but also by programming, along with a computational algorithm, the means for a concurrent analysis, by the machine, of propagation of computational error. We can even arrange a computer program so that the machine itself determines approximation-parameter values in such a way that computational errors are kept smaller than some prescribed tolerance. A great deal of additional sophistication in the combination of analytical, logical, computational, and programming technique can be reasonably expected to be included in the future repertoire of the stored-program computing machine.

One of the things we must do is find various coordinate transformations or changes of variable that we can make before, during, or after a computation in order to obtain more accurate results (or results of greater *known* accuracy) with less effort; ideally these will be transformations which can be performed by the computer. In Chapter 7 we discuss global transformations to be made prior to a computation; primarily, these are based on reducing the order of a system by means of integral surfaces. In Chapter 8 we examine transformations of a local nature, that is, those that change their form in different portions of the solution space; Chapter 9 touches briefly on time-dependent transformations. In Chapter 10 we take a somewhat different tack and look at transformations to be used *after* some calculations have been performed; these methods are of particular importance for error-bounding approaches and for design studies.

7

A PRIORI GLOBAL TRANSFORMATIONS

7.1 INTRODUCTION

We are interested in finding ways to transform a given equation, prior to any computation, in order to make it easier to solve numerically, no matter what approximate method we choose to use, within reason; before doing this we need to know what kinds of equations are difficult to solve and thus in need of transformation. Roughly speaking, difficult problems are those whose solutions may be changing very rapidly over short time intervals; not only does this require that a solution be approximated over short time steps in order to represent the changes, but rapid changes also imply that some derivatives of the solution are large, which means that the local discretization error may well be correspondingly large. Still further, we see that rapidly changing functions are poorly approximated by low-degree polynomials, an approximation technique inherent in most numerical methods. Thus one of the goals of transforming the equations is to straighten out the vector field near a particular solution so that it is more like a uniform translation, in order to eliminate rapid changes in the solution.

Another common difficulty, mentioned before, is that we often seek a specific solution of one type, say a decaying exponential, which is surrounded by other solutions, some of which, like exploding exponentials, are of a radically different nature. In this case the goal of a transformation is, at least in part, to isolate the desired solution from its contaminating neighbors or, if the desired solution is not the dominant one, to make it so.

7.2 SOME SIMPLE CLASSICAL TRANSFORMATIONS

Some simple transformations which help us find a specific solution have been known for many years and can sometimes be of great value. For example, consider the problem in which we need to isolate a physically meaningful decaying solution from nearby growing ones. One simple method here is to integrate backwards in time from infinity. Another method, often more useful, is to use the *Riccati transformation.* Suppose we seek a decaying solution to the problem

$$y'' = a(t)y,$$

where $a(t)$ is a positive, slowly changing function of t. If $a(t)$ were a constant, say $a(t) = \alpha^2$, we would expect solutions of the form

$$e^{\alpha t} \quad \text{and} \quad e^{-\alpha t},$$

the former of which would overwhelm the latter in a numerical computation for many methods. We introduce the new dependent variable $x(t)$ satisfying

$$y' = xy;$$

then we find

$$y'' = xy' + yx' = (x^2 + x')y = a(t)y$$

whence, if $y(t) \neq 0$,

$$x' + x^2 = a(t).$$

Since $x = y'/y$, the solutions $y = e^{\alpha t}$, $y = e^{-\alpha t}$ become

$$x = \alpha, \qquad x = -\alpha,$$

solutions which are well separated from each other; we would expect the numerical solution of this time-dependent, nonlinear, transformed problem to be easier than it was before the transformation. Having found x, we can compute y from

$$y(t) = \exp \int_0^t x(\tau)d\tau.$$

The same ideas can be applied to many kinds of equations, both nonlinear and of higher order.

EXERCISE

Choose a numerical method, and, using a computer, compare numerical solution (accuracy and computing time) of the following example before and after applying the Riccati transformation:

$$y'' = (1 + e^{-(t+10)^2})y, \qquad y(0) = 1, \qquad y(10) = 0.00005.$$

A prototype for the rapidly changing solution is that of the rapidly oscillating one. Since the solution to

$$y'' + \alpha^2 y = 0$$

has the form

$$y = a \sin \alpha t + b \cos \alpha t,$$

which oscillates rapidly for large α, we would expect the same behavior and hence numerical difficulties for the equation

$$y'' + a(t)y = 0,$$

where $a(t)$ is a large, positive, slowly varying function. We try to improve matters via the *Madelung transformation*, which represents y by its amplitude and phase,

$$y(t) = u(t) \cos v(t).$$

Since

$$y' = u' \cos v - uv' \sin v$$

and

$$y'' = u'' \cos v - 2u'v' \sin v - u[v'' \sin v + (v')^2 \cos v],$$

we find

$$\sin v \cdot [-2u'v' - uv''] + \cos v \cdot [u'' - u(v')^2 + a(t)u] = 0.$$

If we try to find a solution such that the two terms in the square brackets vanish, we find

$$u'' - u(v')^2 + a(t)u = 0$$

$$uv'' + 2u'v' = 0.$$

The latter equation integrates immediately to give

$$v' = cu^{-2}, \qquad \text{for some constant } c,$$

so we finally have

$$u'' - c^2u^{-3} + a(t)u = 0$$

to determine the amplitude $u(t)$ where c is to be determined by the given data if possible; the oscillatory behavior has been eliminated. This same method is often of use on more complex nonlinear problems also.

There is no substitute for insight, ingenuity, and plain good luck when it comes to transforming an equation. Since many problems in practice are too complex for Riccati, Madelung, or other classical transformations, other approaches are also needed; a good basic tool is the perturbation approach. For example, if one must solve

$$y' = f(t, y)y,$$

where $f(t, y)$ changes slowly and for large t is nearly the constant A, then writing

$$y = e^{At} + z$$

yields

$$z' = (f(t, e^{At} + z) - A)e^{At} + f(t, e^{At} + z)z,$$

in which both the right-hand side and the solution are expected to be small and slowly varying. Since we cannot possibly enumerate all possible types of special transformations which might be useful for special equations, we conclude this brief discussion of special methods and move to consideration of some more broadly applicable and general classes of transformations.

7.3 IMPLICIT TRANSFORMATIONS OF AUTONOMOUS SYSTEMS

Consider the differential equation

$$\mathbf{y}' = \mathbf{f}(\mathbf{y}), \qquad \mathbf{y} \in E^n;$$

it is often possible to find an implicit transformation from $\mathbf{y}$ to $\mathbf{z}$ in E^n, that is,

$$\mathbf{y} = \mathbf{T}(\mathbf{z}),$$

which yields a simpler or "better behaved" equation for $\mathbf{z}$,

$$\mathbf{z}' = \mathbf{g}(\mathbf{z}).$$

From $\mathbf{y} = \mathbf{T}(\mathbf{z})$, we have

$$\mathbf{y}' = \left(\frac{\partial \mathbf{T}}{\partial \mathbf{z}}\right)\mathbf{z}' = \mathbf{f}(\mathbf{T}(\mathbf{z})),$$

and so

$$\mathbf{z}' = \left(\frac{\partial \mathbf{T}}{\partial \mathbf{z}}\right)^{-1} \mathbf{f}(\mathbf{T}(\mathbf{z}))$$

is the derived equation—assuming, of course, that the matrix $(\partial \mathbf{T}/\partial \mathbf{z})$ is nonsingular. We consider a simple example in some detail.

EXAMPLE Consider the system

$$y_1' = y_2$$

$$y_2' = -y_1.$$

Define a transformation $\mathbf{y} = \mathbf{T}(\mathbf{z})$ by

$$y_1 = z_1 \cos z_2$$

$$y_2 = z_1 \sin z_2,$$

with $\mathbf{y} = \begin{pmatrix} y_1 \\ y_2 \end{pmatrix}$ and $\mathbf{z} = \begin{pmatrix} z_1 \\ z_2 \end{pmatrix}$; we motivate this choice later. Then

$$\left(\frac{\partial \mathbf{T}}{\partial \mathbf{z}}\right) = \begin{pmatrix} \cos z_2 & -z_1 \sin z_2 \\ \sin z_2 & z_1 \cos z_2 \end{pmatrix}$$

and

$$\left(\frac{\partial \mathbf{T}}{\partial \mathbf{z}}\right)^{-1} = \begin{pmatrix} \cos z_2 & \sin z_2 \\ \dfrac{-\sin z_2}{z_1} & \dfrac{\cos z_2}{z_1} \end{pmatrix} \qquad (\text{if } z_1 \neq 0)$$

and therefore

$$\begin{aligned} z_1' &= (\cos z_2)(z_1 \sin z_2) + (\sin z_2)(-z_1 \cos z_2) \\ &= 0, \end{aligned}$$

$$z_2' = -\left(\frac{\sin z_2}{z_1}\right)(z_1 \sin z_2) + \left(\frac{\cos z_2}{z_1}\right)(-z_1 \cos z_2)$$

$$= -1.$$

The derived differential system in the variables z_1, z_2 is thus

$$z_1' = 0$$

$$z_2' = -1.$$

Any reasonable computational method should give essentially exact results for *this* system! The solutions are $z_1(t) = z_1(0)$, $z_2(t) = z_2(0) - t$. Applying the transformation $\mathbf{y} = \mathbf{T}(\mathbf{z})$, we then have

$$y_1(t) = z_1(0) \cos (z_2(0) - t)$$

$$y_2(t) = z_1(0) \sin (z_2(0) - t).$$

For an initial value problem we have

$$y_1(0) = z_1(0) \cos z_2(0)$$

$$y_2(0) = z_1(0) \sin z_2(0);$$

given $y_1(0)$, $y_2(0)$, corresponding initial values of z_1 and z_2 can be found approximating

$$z_1(0) = (y_1^2(0) + y_2^2(0))^{1/2}$$

$$z_2(0) = \cos^{-1}\{y_1(0)/(y_1^2(0) + y_2^2(0))^{1/2}\}.$$

The value $z_2(0)$ is determined only modulo 2π. The coordinate z_2 is a "periodic" coordinate for this problem. The transformation $\mathbf{y} \to \mathbf{z}$ just discussed is, of course, a change from rectangular to "polar" coordinates.

For a boundary value problem described, say, by boundary conditions

$$\varphi_1(y_1(0), y_2(0)) = 0$$

$$\varphi_2(y_1(T), y_2(T)) = 0,$$

the corresponding boundary conditions in the variables z_1, z_2 are

$$\varphi_1(z_1(0) \cos z_2(0), z_1(0) \sin z_2(0)) = 0$$

$$\varphi_2(z_1(T) \cos z_2(T), z_1(T) \sin z_2(T)) = 0.$$

Making use of the expression for the *solution* of the differential equations for

z_1, z_2 passing through $z_1(0)$, $z_2(0)$ and given by $z_1(t) = z_1(0)$, $z_2(t) = z_2(0) - t$ we can put the boundary conditions in the form

$$\varphi_1(z_1(0) \cos z_2(0), z_1(0) \sin z_2(0)) = 0$$

$$\varphi_2(z_1(0) \cos (z_2(0) - T), z_1(0) \sin (z_2(0) - T)) = 0.$$

In the special case that $\varphi_1 = 0$ and $\varphi_2 = 0$ are of the form $y_1(0) - a = 0$, $y_1(T) - b = 0$ for some given constants a and b, the corresponding boundary conditions in the z variables become

$$z_1(0) \cos z_2(0) - a = 0$$

$$z_1(T) \cos z_2(T) - b = 0.$$

It is interesting to compare the geometric interpretations of these two formulations of the boundary value problem. In **y** coordinates, the integral curves in the phase plane (y_1, y_2 plane) are concentric circles about the origin, and the flow consists of a "rigid" rotation of the phase space about the origin with a constant angular velocity of 1. See Figure 7–1.

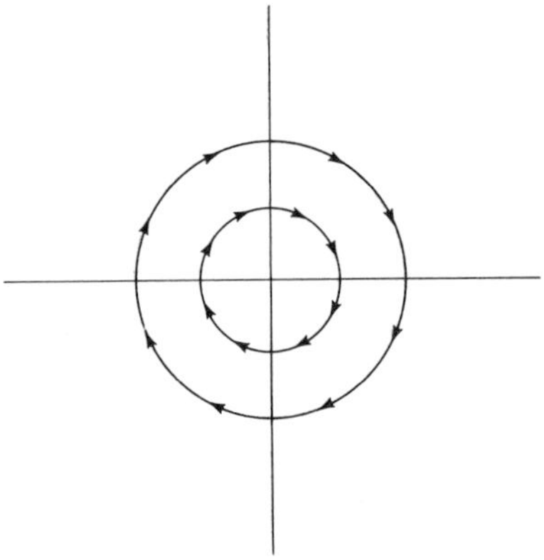

FIGURE 7–1

Suppose $0 < a < b$; then the boundary value problem

$$y_1' = y_2$$

$$y_2' = -y_1$$

$$y_1(0) - a = 0$$

$$y_1(T) - b = 0$$

has the geometrical significance indicated in Figure 7–2.

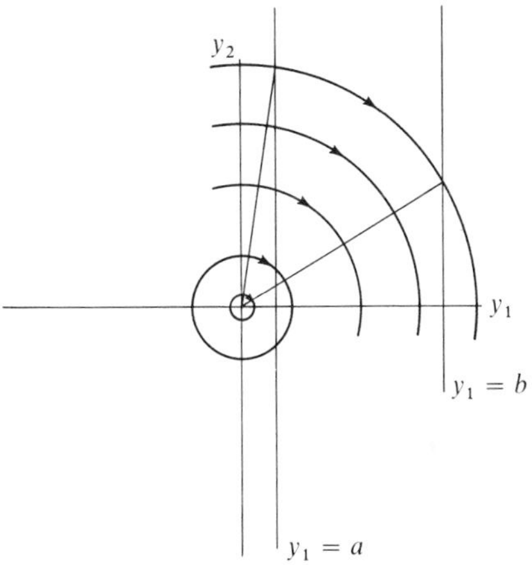

FIGURE 7–2

The problem consists geometrically of finding *an arc* of one of the circles with end points on the lines $y_1 = a$ and $y_1 = b$ which subtends an angle of T radians at the origin. To do this we must find y_2 such that

$$\frac{\tan^{-1} y_2}{a - \tan^{-1}} \sqrt{\frac{a^2 - b^2 + y_2^2}{b}} = T.$$

On the other hand, in $\mathbf{z}$ coordinates the integral curves are straight lines and the flow consists of a rigid parallel translation of the phase space in the direction of the negative z_2-axis with a constant velocity of 1. See Figure 7–3.

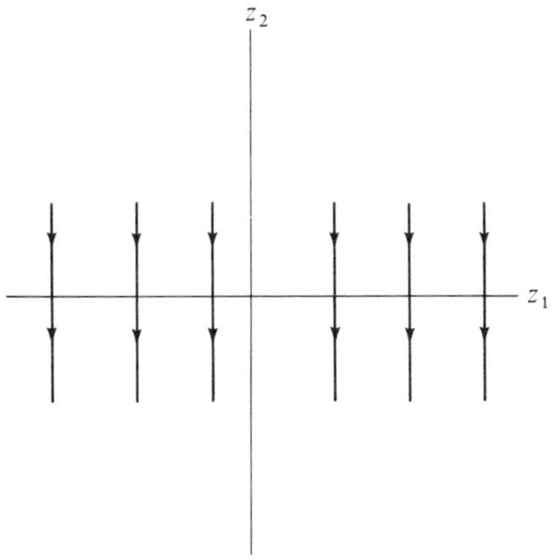

FIGURE 7–3

The same boundary value problem just described has the form

$$z_1' = 0$$

$$z_2' = -1$$

$$z_1(0) \cos z_2(0) - a = 0$$

$$z_1(T) \cos z_2(T) - b = 0$$

in **z** coordinates and has the geometrical significance indicated in Figure 7–4.

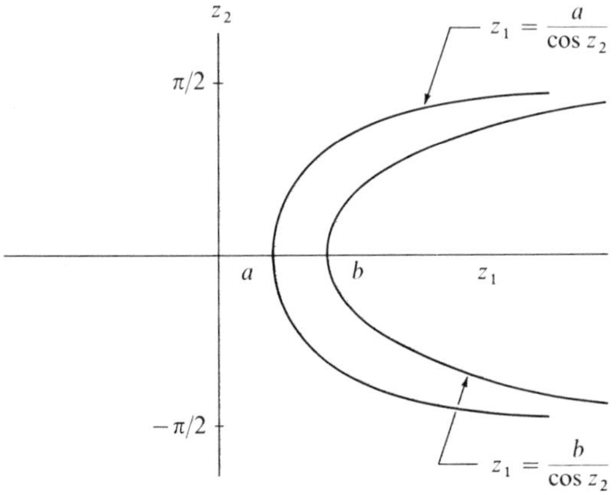

FIGURE 7–4

The problem consists geometrically in finding *a segment* of one of the straight lines with end points on the curves $z_1 = a/\cos z_2$ and $z_1 = b/\cos z_2$ of length T. To do this we must find z_2 such that

$$\frac{b-a}{\cos z_2} = T.$$

This completes our study of the example. ●

EXERCISE

Choose a numerical method and compare its application to the two different formulations of the boundary value problem just discussed. Compare computing times required for nearly equal accuracy.

Let us return now to the beginning of the discussion of this example and ask: "How did we choose the transformation $\mathbf{y} = \mathbf{T}(\mathbf{z})$ in the first place?" Our choice of $y_1 = z_1 \cos z_2$, $y_2 = z_1 \sin z_2$, was based, of course, on (1) our knowledge that integral curves are circles in the y_1, y_2 plane with center at the origin, and (2) a convenient parametrization of these circles using "polar" coordinates.

In order to see the general approach, let us consider next a slightly more complicated example, namely the equations for the nonlinear pendulum discussed in Section 2.7.

EXAMPLE Consider the system

$$y_1' = y_2$$

$$y_2' = -\sin y_1$$

and the integral curves

$$\tfrac{1}{2}\, y_2^2 - \cos y_1 = \text{constant}$$

(see Figure 2–2 and Figure 2–3). The integral curve $\frac{1}{2}y_2^2 - \cos y_1 = C$ is nearly a circle for C slightly greater than -1 (actually infinitely many circles, one each about the points $\ldots, -2\pi, 0, 2\pi, 4\pi, \ldots$). For C greater than $+1$ the integral curve $\frac{1}{2}y_2^2 - \cos y_1 = C$ is a wavy line which is nearly a straight line for very large C (actually a *pair* of nearly straight lines).

If we put $z_1 = \frac{1}{2}y_2^2 - \cos y_1$, then a value of z_1 puts us on one of the integral curves. For $z_1 > 1$, $\frac{1}{2}y_2^2 - \cos y_1 = z_1$ implies (y_1, y_2) is on one of the wavy lines, whereas for $-1 < z < +1$, any (y_1, y_2) satisfying $\frac{1}{2}y_2^2 - \cos y_1 = z_1$ is on one of the near-circles (see Figure 7–5).

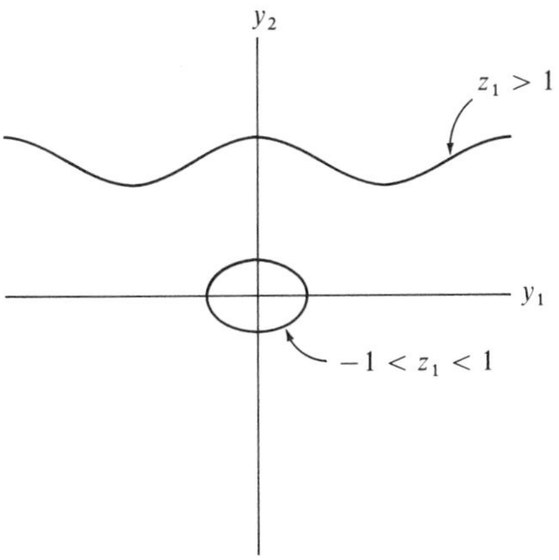

FIGURE 7–5

We would like to choose a second coordinate, z_2, pinning down a particular point on one of these curves. For the near-circle we might use an angular variable for z_2, that is, a periodic coordinate, say defined by

$$\tan z_2 = \frac{y_2}{y_1}.$$

This would lead to the derived system

$$z_1' = 0$$

$$z_2' = \frac{1}{1+\tan^2 z_2}\left(-\tan^2 z_2 - \frac{\sin y_1}{y_1}\right),$$

where y_1 can be determined from z_1, z_2 by solving

$$z_1 = \tfrac{1}{2}y_2^2 - \cos y_1$$

$$\tan z_2 = \frac{y_2}{y_1}.$$

In fact we find, *for* z_1 *near* -1 and using

$$\frac{\sin y_1}{y_1} = 1 - \frac{y_1^2}{3!} + \frac{y_1^4}{5!} - \dots$$

and

$$\tfrac{1}{2}y_1^2 \tan^2 z_2 = z_1 + \cos y_1$$

$$= z_1 + 1 - \frac{y_1^2}{2!} + \frac{y_1^4}{4!} - \dots,$$

that

$$y_1^2 = 2\left(\frac{z_1+1}{1+\tan^2 z_2}\right)\left(1 + \frac{z_1+1}{6(1+\tan^2 z_2)^2} + \dots\right),$$

and so

$$\frac{\sin y_1}{y_1} = 1 - \tfrac{1}{3}\,\frac{(z_1+1)}{1+\tan^2 z_2} + O((z_1+1)^2).$$

The derived system thus has the form

$$z_1' = 0$$

$$z_2' = -1 + \tfrac{1}{3}\,\frac{z_1+1}{(1+\tan^2 z_2)^2} + O((z_1+1)^2)$$

and has for z_1 near -1 the appearance of a perturbation of the system

$$z_1' = 0$$

$$z_2' = -1$$

derived for the previous example.

For z_1 larger than $+1$, we could choose as a parameterization of a point on the wavy line $\frac{1}{2}y_2^2 - \cos y_1 = z_1$, the coordinate $z_2 = y_1$. In this case the derived system becomes

$$z_1' = 0$$

$$z_2' = \sqrt{2(z_1 + \cos z_2)};$$

and for very large z_1, this is approximately

$$z_1' = 0$$

$$z_2' = \sqrt{2z_1} = \text{constant}.$$

It is not clear from our discussion of the cases in which z_1 is near -1 or is much larger than $+1$ that there is any advantage computationally in using the indicated transformations; the discussion merely suggests that there might well be.

EXERCISE

Try some numerical method and see!

In addition to the nearly constant behavior of the right-hand sides of the z_2' equations in the cases discussed there is another important feature of the derived system in the **z** variables resulting from the choice of the value of the *integral* function as the variable $z_1 = \frac{1}{2}y_2^2 - \cos y_1$. It is, of course, that $z_1' = 0$; that is, z_1 remains constant along the solution. The transformation has, in effect, reduced the system from second order to first order. ●

Additional reference: 80.

7.4 IMPLICIT TRANSFORMATIONS: A NUMERICAL EXAMPLE

The final example of the preceding section was chosen because its geometry indicates quite nicely the techniques one can use to discover useful transformations; as we remarked in that section, however, it is not clear that transformations are valuable for that problem. Consider for example the initial value problem with

$$y_1(0) = 0, \qquad y_2(0) = 20$$

for which the integral-surface constant C equals 199. For a value of C that is this large, the integral-surface wavy line is very nearly straight, indicating that y_2 varies only slightly from its initial value of 20, which indicates in turn that $y_1(t)$ is roughly equal to $20t$; the transformation to the z_1, z_2 equation states this precisely but does not seem to be really useful since the solution to the original problem looks so well behaved. Some numerical experiments were performed, solving this problem both in the original and in the transformed form. With a fixed-step-size second-order Runge-Kutta formula for the interval $0 \leq t \leq 3$, the solution via the transformed equation was invariably about four times as accurate (relative error) as that obtained directly, but the computation required about $\frac{7}{3}$ as much time because of the transformation back to y. To obtain the same accuracy, almost the same time was required for each method. For long periods of computation this would give a very slight edge to the transformed method since it can achieve a given accuracy in half as many steps, thus somewhat reducing rounding error effects. The basic conclusion, however, is that our intuition was correct; the solution is not varying rapidly enough to make transformations useful.

We shall now discuss an example for which both intuition and numerical experiment indicate that a transformation is of value. In many kinds of problems the solution varies rapidly at first and then settles down; this usually requires a very short step size to represent the solution. We shall show how a transformation can be used to help with this problem also.

Consider the differential equation

$$\begin{aligned} y_1' &= e^{y_2} \\ y_2' &= y_1 \end{aligned} \tag{7–1}$$

with initial conditions $y_1(0) = -100$, $y_2(0) = 8.50713$. It is easy to find an integral for this system, namely

$$\frac{y_1^2}{2} - e^{y_2} = C = \frac{(100)^2}{2} - e^{8.50713} \sim 50.$$

The integral curve is shown in Figure 7–6 for $y_1 < -10$ (since $y_1^2/2 = C + e^{y_2} > C = 50$ implies $|y_1| > 10$).

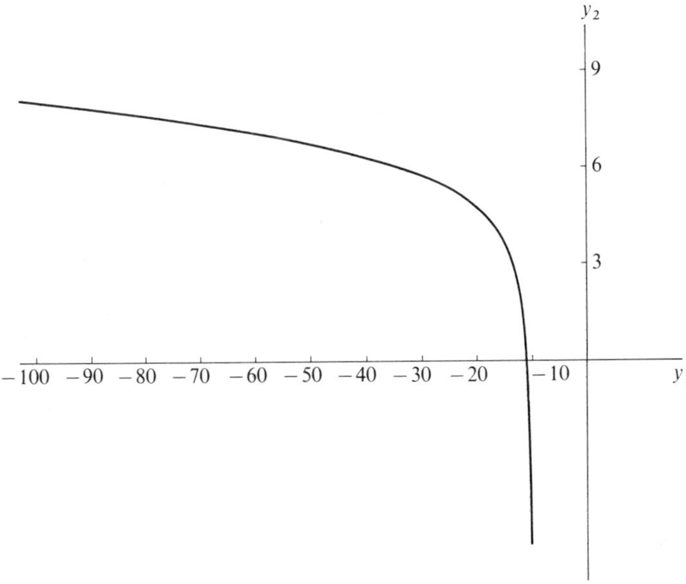

FIGURE 7–6

For the values of y_1 near -10, small changes in y_1 correspond to large changes in y_2, whereas the situation is reversed for y_1 nearer -100. From the differential equation it is clear that y_1 always increases with t but is bounded above by -10, which implies in this case that y_1' tends to zero; that is, y_2 tends to $-\infty$ and y_1 tends to -10. It is apparent that y_1 changes very rapidly at first, while y_2 does not, but that y_1 varies slowly later, while y_2 explodes towards $-\infty$. Thus we expect great difficulties in computing the solution for all time values, since one of the components of the solution always changes rapidly. Because of the integral surface relationship between y_1 and y_2, however, we can deduce differential equations for y_1 and y_2 *separately*, that is

$$y_1' = \frac{y_1^2}{2} - 50,$$

$$y_2' = -\sqrt{100 + 2e^{y_2}}.$$

The second of these equations is relatively easy to solve numerically for $8.5 \geq y_2 \geq 2.4$, and the first is easy thereafter where $-11 \leq y_1 \leq -10$. In essence this suggests two different transformations from (y_1, y_2) to (z_1, z_2). For $8.5 \geq y_2 \geq 2.4$ we take

$$z_1 = \frac{y_1^2}{2} - e^{y_2}, \qquad z_2 = y_1,$$

so that

$$z_1' = 0, \qquad z_2' = \frac{z_2^2}{2} - 50,$$

and for $2.4 \geq y_2 \geq -\infty$ we take

$$z_1 = \frac{y_1^2}{2} - e^{y_2}, \qquad z_2 = y_2,$$

so that

$$z_1' = 0, \qquad z_2' = -\sqrt{100 + 2e^{z_2}}.$$

Mathematically, of course, we have a single transformation, but intuitively we have two; we will return to this in the next section.

EXERCISE

Compare numerical solutions obtained by using the same numerical method on the original and transformed equations.

For illustrative purposes, let us make things intentionally harder for ourselves; let us try to solve

$$y' = \frac{y^2}{2} - 50, \qquad y(0) = -100 \tag{7-2}$$

numerically for small time, where solution is most difficult. We seek a transformation from y to z which parametrizes the curve in Figure 7–6 so that z does not vary so rapidly. From the shape of the curve, we are led to try

$$y = -10 - ae^z, \qquad a > 0 \text{ constant.}$$

Substituting into the differential equation yields

$$z' = \frac{-a}{2} e^z - 10,$$

which indicates that $z(t)$ contains a term like $-10t$. If we take advantage of this and let

$$z = 10(w - t),$$

we find

$$w' = \frac{-a}{20}\,e^{10(w-t)},$$

$$y(t) = -10 - ae^{10(w-t)}.$$

If we let $a = 90$ we get $w(0) = 0$; the final equation after these transformations is

$$\begin{aligned} w' &= -4.5e^{10(w-t)}, \qquad w(0) = 0, \\ y(t) &= -10 - 90e^{10(w-t)}. \end{aligned} \tag{7-3}$$

We have performed some simple numerical experiments to see whether or not the transformation to Eq. 7–3 makes it easier to solve Eq. 7–2 for small t, the difficult area. Consider, for example, solving the two equations by a second-order Runge-Kutta formula with fixed step size h. For $h = 0.05$, the solution to Eq. 7–3 yields a solution $y(t)$ for Eq. 7–2 with relative error of at most 13% for grid points in $0 \leq t \leq .2$, over which region the true solution changes from -100 to -12.5; although this is not very accurate, we remark that using $h = 0.05$ on Eq. 7–2 yields a relative error of 3×10^5 in this region. For $h = 0.01$, a value for which we can hope to represent the solution more accurately, dealing with $w(t)$ yields a solution that is more than twice as accurate as solving Eq. 7–2 directly, with an additional time cost of only 33% for the transformation back to y. By further reducing the step size for the direct approach in order to get accuracy comparable to that obtained by the transformed equation with $h = 0.01$, we find that we spend about 15% more time by using the direct method. Thus, although for small periods of time these comparisons may not be vital, they indicate that transformations can reduce the computing time and even, in some cases, make a numerical method usable when it would not be so without the transformation.

EXERCISE

Try the fourth-order Runge-Kutta method on the original and transformed equations. Compare results.

7.5 CHANGING THE FORM OF A TRANSFORMATION

In the preceding sections we have considered transformations based on the use of integral surfaces $H(y_1, \ldots, y_n) = \text{constant}$; we saw that this was in reality the same as using the integral surface relationship to solve for one variable, say y_1, in terms of the others and thus reduce the order of the differential equation. Unfortunately, $H(y_1, \ldots, y_n) = C$ does not usually define y_1 as a single-valued function of $y_2, \ldots, y_n$ meaning that branch points of the solution

$$y_1 = g(y_2, \ldots, y_n)$$

may occur. We circumvent this by solving for different components $y_1, \ldots, y_n$ in different parts of the trajectory.

Consider again the equations for a nonlinear spring or pendulum, which were discussed extensively in Section 7.3,

$$y_1' = y_2$$

$$y_2' = -\sin y_1.$$

This system has the integral

$$\frac{y_2^2}{2} - \cos y_1 = C;$$

solving for y_2 yields

$$y_2 = \{2(C + \cos y_1)\}^{1/2}.$$

If we are interested in the initial value problem with $y_1(0)$ and $y_2(0)$ given and, say, $y_2(0) > 0$, $y_1(0) = 0$, then

$$C = \tfrac{1}{2}y_2^2(0) - \cos 0 = \tfrac{1}{2}y_2^2(0) - 1,$$

so that

$$y_2 = (y_2^2(0) - 2 + 2\cos y_1)^{1/2}. \tag{7–4}$$

Since $z_1 = C$ in Figure 7–5, we can see from the figure that for $C > 1$, that is, for $y_2(0) > 2$, Eq. 7–4 defines y_2 as a single-valued function of y_1 for *all* y_1 and the system

$$y_1' = y_2,$$

$$y_2' = -\sin y_1$$

is reduced to

$$y_1' = (y_2^2(0) - 2 + 2\cos y_1)^{1/2}. \tag{7–5}$$

If we find $y_1(t)$ satisfying Eq. 7–5, say *approximately*, then $y_2(t)$ can be determined from Eq. 7–4 as accurately and as easily as we can find a value of the cosine and square-root functions for the given arguments.

On the other hand, again viewing Figure 7–5 for $y_1(0) = 0$ and $0 < y_2 < 2$, we can see a difficulty arising when y_1 reaches the value for which $y_2 = 0$, where the square root in Eq. 7–5 has a branch-point singularity. In Figure 7–7 we illustrate a way around this difficulty, namely to solve alternately the integral

$$\tfrac{1}{2}y_2^2 - \cos y_1 = \tfrac{1}{2}y_2^2(0) - 1$$

for y_2 or y_1 in the indicated regions.

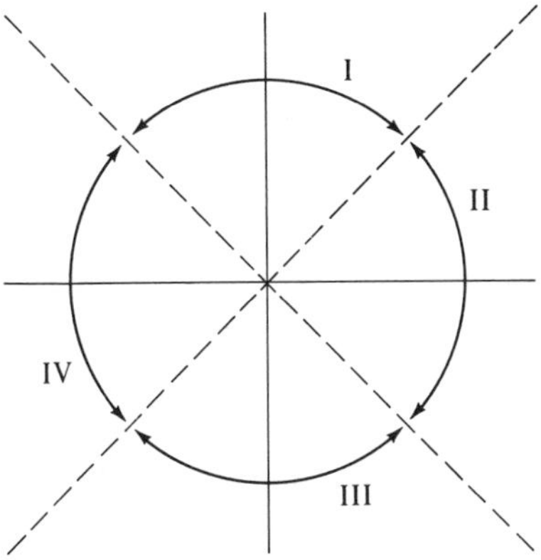

FIGURE 7–7

In I, we can solve for

$$y_2 = (y_2^2(0) - 2 + 2\cos y_1)^{1/2} > 0;$$

in II, we can solve for

$$y_1 = \cos^{-1}\{\tfrac{1}{2}y_2^2 - \tfrac{1}{2}y_2^2(0) + 1\} \qquad \text{with } 0 < y_1 < \pi;$$

in III, we can solve for

$$y_2 = -(y_2^2(0) - 2 + 2\cos y_1)^{1/2} < 0;$$

and in IV, we can solve for

$$y_1 = -\cos^{-1}\{\tfrac{1}{2}y_2^2 - \tfrac{1}{2}y_2^2(0) + 1\} \qquad \text{with } -\pi < y_1 < 0.$$

In this way we can reduce the given second-order system to a different first-order equation in each of the four regions by using the known integral.

In the next chapter we will consider a generalization of this procedure to integrals that are entirely local.

EXERCISE

Choose a numerical method and compare results for the original and transformed equations.

Additional reference: 80.

7.6 INTEGRALS IN CELESTIAL MECHANICS

In the previous sections we have considered the use of integrals to simplify computations—for example, by reducing the order of a system. At this point, we shall pause and digress somewhat from our narrative about transformations to indicate the use that has been made of integrals, for transformations and other purposes, in the extremely complicated subject of celestial mechanics: an example of how theoretical, qualitative concepts can be used to help obtain quantitative results.

In Section 1.3 the differential equations of motion for the two-body problem were given along with two integrals. The system is of fourth order and the phase space is E^4. In the two-dimensional position space, the solution curves (in E^4) have projections that are conic sections, as has been known since the studies of Kepler and Newton. In the velocity space (or hodograph plane) the solution curves have projections which lie on circles.

These facts are widely used in celestial mechanics and have recently been applied to the machine computation of orbits of artificial satellites and trajectories of space probes. An orbiting satellite, for example, can as a first approximation be viewed as moving in some fixed plane on a Kepler ellipse with the center of the earth at one focus. In order to improve this approximation a common practice is to consider a parameterization of the solution in terms of, say, eccentricity, semi-major axis, the inclination of the ellipse in its plane, the orientation (or position) of the plane of motion itself, and the angular position of the satellite on the ellipse. If these quantities are allowed to vary with time, then differential equations for their time derivatives are derived that take into account various perturbing forces, such as oblateness of the earth, attraction of the sun and moon, "drag" of the earth's atmosphere, and so on. We will not attempt a catalog of the vast literature on this subject here. The interested reader should not have much trouble finding his way into the midst of it in a scientific or engineering library.

In a rotating coordinate system, with one axis through the two large bodies, the differential equations describing the restricted three-body problem can be written

$$
\begin{aligned}
y_1'' &= y_1 + 2y_2' - \frac{(1-\mu)(y_1+\mu)}{r_1^3} - \frac{\mu(y_1 - 1 + \mu)}{r_2^3} \\
y_2'' &= y_2 - 2y_1' - \frac{(1-\mu)y_2}{r_1^3} - \frac{\mu y_2}{r_2^3}
\end{aligned}
\tag{7–6}
$$

with

$$
\begin{aligned}
r_1 &= ((y_1+\mu)^2 + y_2^2)^{1/2} \\
r_2 &= ((y_1 - 1 + \mu)^2 + y_2^2)^{1/2}.
\end{aligned}
$$

These equations describe the motion in the y_1, y_2 plane of a small third body that is gravitationally attracted by two large bodies which are themselves rotating in circles about their common center of mass (at the origin) and whose mass ratio is $m_2/(m_1 + m_2) = \mu$ (Figure 7–8).

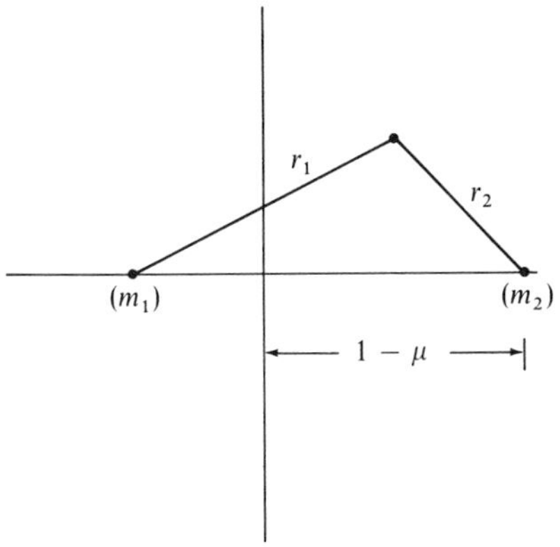

FIGURE 7–8

The units of length, mass, and time are chosen so that the distance between m_1 and m_2 is 1, the sum of the masses is 1, and the gravitational constant is 1. Where m_1 is the earth and m_2 the moon, for example, the ratio of masses is approximately $m_2/(m_1 + m_2) = \mu = 0.01215$.

Of course, the earth-moon-spaceship system is only approximately described by the restricted three-body problem even when these three are assumed to move in a fixed plane. The equations have been of interest in celestial mechanics as a model, for example, of the sun-planet-satellite system, sun-earth-moon and sun-Jupiter-moon systems, as well as the sun-Jupiter-asteroid system.

There are five rest points (where the small third body will remain at rest in the rotating co-ordinate system) denoted by L_1, L_2, L_3, L_4, L_5 in the restricted three-body problem. Two are at the vertices of the two equilateral triangles with base consisting of the line segment $\overline{m_1 m_2}$. These are called the Lagrange solutions, after their discoverer. The other three rest points are on the y_1 axis at positions: (1) to the left of m_1, (2) between m_1 and m_2 and (3) to the right of m_2; the distances depend on μ. See Figure 7–9.

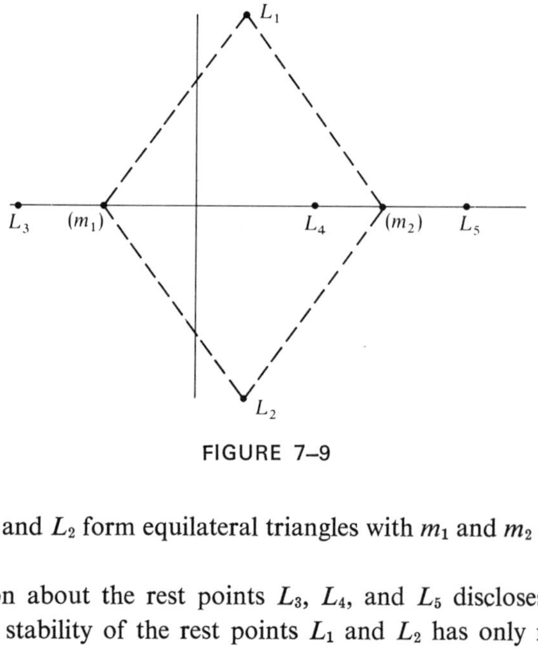

FIGURE 7–9

The points L_1 and L_2 form equilateral triangles with m_1 and m_2 *independently* of μ.

Linearization about the rest points L_3, L_4, and L_5 discloses them to be unstable. The stability of the rest points L_1 and L_2 has only recently been completely settled and depends on the mass ratio μ. If $\mu < (1/27)$, these rest points are stable. For an introductory discussion of these and other questions concerning this problem see [108].

The equations have an integral, known as the Jacobi integral,

$$J(y_1, y_2, y_1', y_2') = \tfrac{1}{2}(y_1'^2 + y_2'^2 - y_1^2 - y_2^2) - \frac{1-\mu}{r_1} - \frac{\mu}{r_2}. \qquad (7\text{–}7)$$

EXERCISE

Show that this *is* an integral of Eq. 7-6.

Every solution lies on a three-dimensional surface, $J =$ constant, in the four-dimensional phase space. The system can be reduced from fourth to third order by solving $J =$ constant for one of the variables y_1, y_2, y_1', y_2' as a function of the other three over a part of a trajectory for which the chosen variable is monotonic, increasing or decreasing with the independent variable t. The integral can also be used as a check on (or measure of) the accuracy of an approximate solution to an initial value problem obtained using the fourth order system as it stands. Starting at an initial point $y_0 = (y_{10}, y_{20}, y_{10}', y_{20}')$, the solution $\mathbf{y}(t, \mathbf{y}_0)$ satisfies

$$J(y_1(t, \mathbf{y}_0), y_2(t, \mathbf{y}_0), y_1'(t, \mathbf{y}_0), y_2'(t, \mathbf{y}_0)) = J(y_{10}, y_{20}, y_{10}', y_{20}') = J_0$$

so the difference

$$J(Y_1(t, \mathbf{y}_0), Y_2(t, \mathbf{y}_0), Y_3(t, \mathbf{y}_0), Y_4(t, \mathbf{y}_0)) - J_0$$

can be taken as a measure of the accuracy of an approximate solution $\mathbf{Y}(t, \mathbf{y}_0)$.

Topological characterization of the integral manifolds $J =$ constant has been given by G. D. Birkhoff and depends on the value of the constant. There are five different cases. For example, for large negative values of the constant, the surface $J =$ constant, with J given by Eq. 7–7, is topologically equivalent to projective three-space, i.e., a solid three-sphere with diametrically opposite points identified. If we put $y_1' = y_2' = 0$ in the equation $J = c$ the resulting curves

$$-\tfrac{1}{2}(y_1^2 + y_2^2) - \frac{1-\mu}{r_1} - \frac{\mu}{r_2} = c \qquad (7\text{–}8)$$

(called *Hill's* curves) separate the (y_1, y_2) plane into regions of possible motion and regions where no solution exists.

EXERCISE

Why can't a solution cross these Hill's curves?

For example, for large negative values of c the curves look as shown in Figure 7–10.

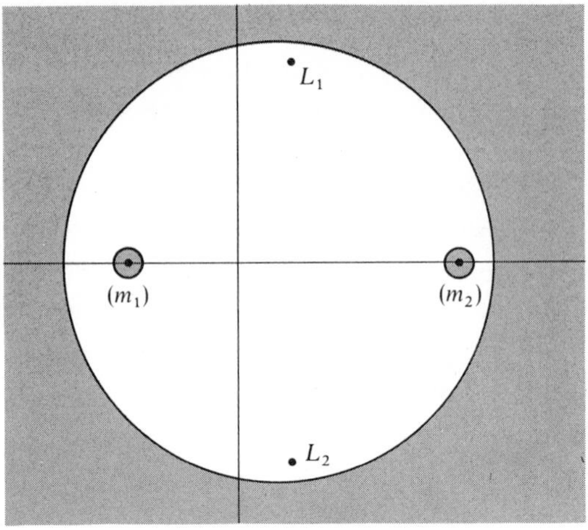

FIGURE 7–10

The regions of possible motion, lying on $J = c$, have projections in the y_1, y_2 plane lying *inside* the two small ovals about m_1 and m_2 and *outside* a larger oval containing m_1, m_2, L_1 and L_2. For a *real* solution, we must have $y_1'^2 + y_2'^2 \geq 0$. It can be seen from Eq. 7–7 and Eq. 7–8 that (y_1, y_2) must lie in the shaded regions indicated in Figure 7–10 in order to have $J = c$, or

$$-\tfrac{1}{2}(y_1^2 + y_2^2) - \frac{1-\mu}{r_1} - \frac{\mu}{r_2} = c - \tfrac{1}{2}(y_1'^2 + y_2'^2)$$

for large negative c.

EXERCISE

Examine the Hill's curves in Figure 7-10 for large negative c. Verify their qualitative appearances.

For example, if we start with an initial $\mathbf{y}_0 = (y_{10}, y_{20})$ near m_1 with a velocity $\mathbf{y}_0' = (y_{10}', y_{20}')$ such that $J(y_{10}, y_{20}, y_{10}', y_{20}')$ is a large negative number, then the solution $\mathbf{y}(t, \mathbf{y}_0, \mathbf{y}_0')$ will remain within a small distance of m_1 for all t (the case of a satellite of m_1).

For $c < c_1$ (a negative number depending on μ), the picture will remain qualitatively (topologically) the same as in Figure 7–10. For $c_1 < c < c_2$, the picture changes and the regions of possible motion in the (y_1, y_2) plane look like those shaded in Figure 7–11. The two small ovals have grown and coalesced.

EXERCISE

Determine c_1 and c_2 for $\mu = \frac{1}{2}$.

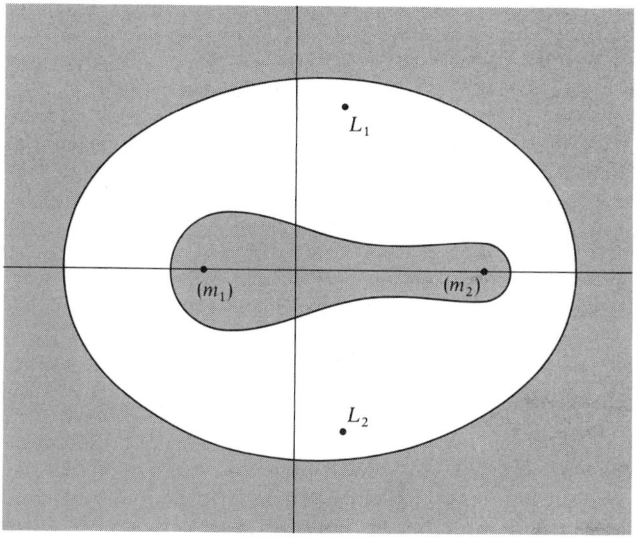

FIGURE 7–11

In this case, transit between neighborhoods of m_1 and m_2 is possible. In fact, it has been shown [28] that there are trajectories which, starting near m_1, approach m_2, loop around m_2 (N times), return toward m_1, loop around m_1 (M times), and then do any number of other similar stunts.

For $c_2 < c < c_3$, another picture develops. This time an escape from the vicinity of both m_1 and m_2 is possible. The only *excluded* region now is the unshaded part of Figure 7–12.

EXERCISE

Determine c_3 for $\mu = \frac{1}{2}$.

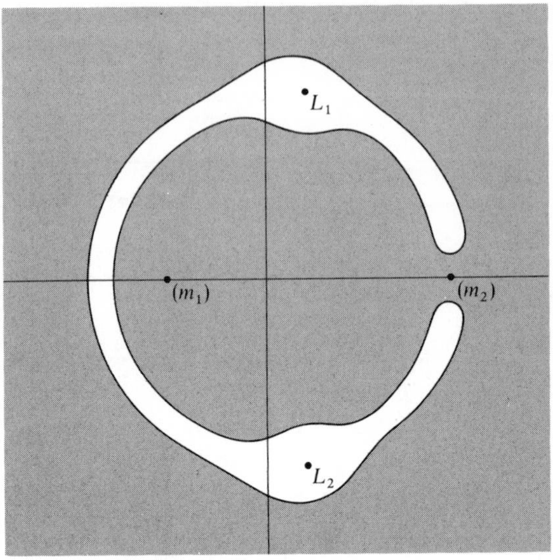

FIGURE 7–12

For still increasing c, the picture for $c_3 < c < c_4$ changes to that indicated in Figure 7–13. The only *excluded* regions are now the interiors of two ovals about L_1 and L_2.

EXERCISE

Determine c_4 for $\mu = \frac{1}{2}$.

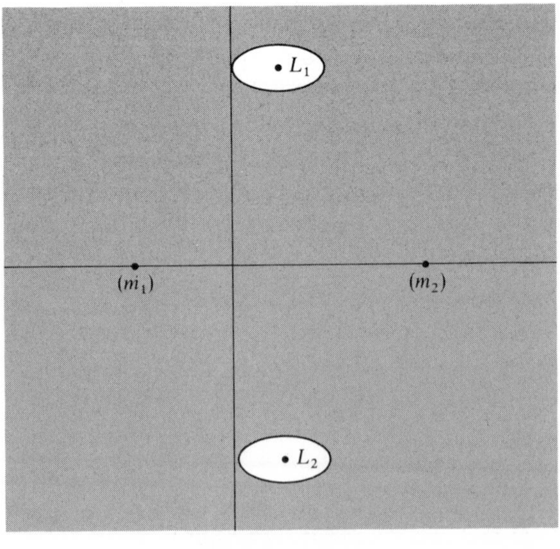

FIGURE 7–13

For $c > c_4$ there are *no* excluded parts of the (y_1, y_2) plane and a trajectory exists starting with any initial position y_{10}, y_{20}, such that

$$J(y_{10}, y_{20}, y'_{10}, y'_{20}) = c > c_4.$$

The value c_4 can be found by putting $r_1 = r_2 = 1$ and $y_1 = (-\mu + 1 - \mu)/2 = \frac{1}{2} - \mu$ and $y_2 = \frac{1}{2}\sqrt{3}$ in Eq. 7–8 to obtain

$$\begin{aligned} c_4 &= -\tfrac{1}{2}((\tfrac{1}{2} - \mu)^2 + \tfrac{3}{4}) - 1 + \mu - \mu \\ &= -\tfrac{1}{2}(3 - \mu + \mu^2). \end{aligned}$$

To remark that the trajectories for the restricted three-body problem are of a greater and more complicated variety than the conics of the two-body problem would be an understatement. In Figure 7–14 we give a rough sketch of only a few of the various kinds of *periodic* solutions found [97] for the case $\mu = \frac{1}{2}$, that is, equal masses.

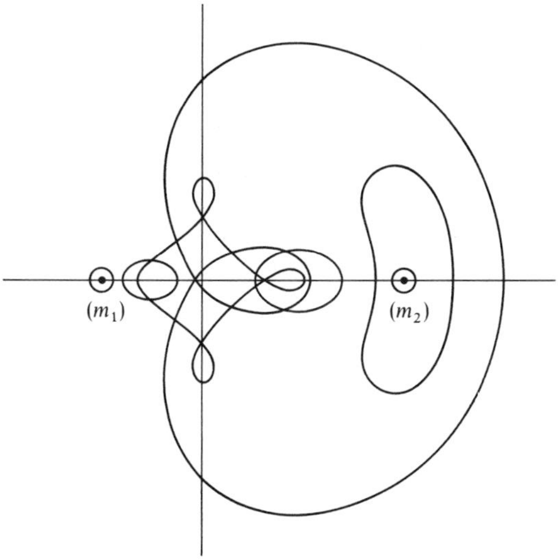

FIGURE 7–14

A solution for the mass ratio 1/81.375 corresponding to the earth-moon system has been computed [12], and looks like that shown in Figure 7–15.

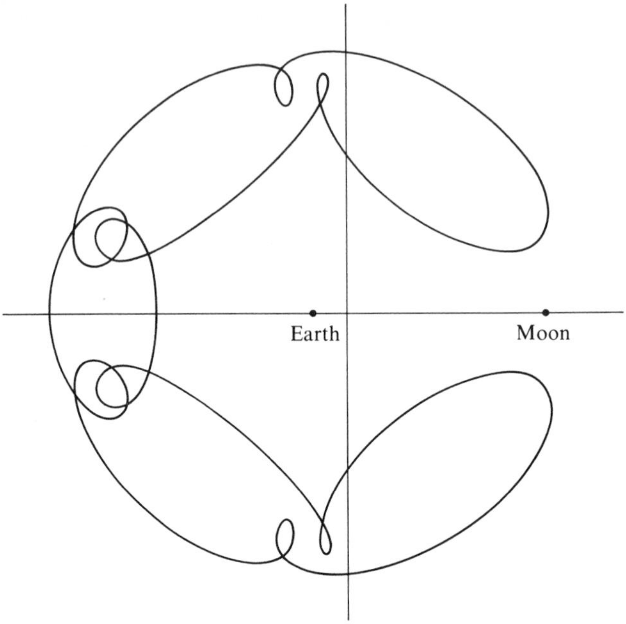

FIGURE 7–15

EXERCISE

Using the restricted three-body equations with $\mu = 1/81.375$ as a model of earth-moon-spaceship (and a computer for the numerical work, of course!), choose a numerical method (for example, fourth-order Runge-Kutta), compute and plot some trajectories which begin near the earth and pass near the moon. Study the *times* of earth-to-moon transit of the spaceship for initial velocities near minimum "escape" (from earth) velocity.

Additional reference: 108.

7.7 REGULARIZATION

For the purpose of studying the nearly singular problems in celestial mechanics in which the small third body almost collides or does collide with one of the large bodies, special transformations have been developed which *regularize* singularities of the problem. Extensive and significant studies of the restricted three-body problem above were made [97] using complex regularization transformations for near-collision trajectories; in this way numerical methods were combined with subtle analytical techniques to obtain a thorough picture of a large number of families of trajectories. We shall give an example of regularization applied to the simpler two-body problem mentioned in Section 1.3.

Consider the equations

$$x_1'' = \frac{-x_1}{r^3}$$

$$x_2'' = \frac{-x_2}{r^3},$$

where

$$r = (x_1^2 + x_2^2)^{1/2}.$$

By regularization at the singularity $r = 0$ of the vector field it can be shown that solutions can be continued analytically for all time and are unique for any initial point other than the origin $r = 0$.

For solutions through initial points on the constant angular-momentum surface $x_1x_2' - x_2x_1' = 0$ we have

$$\frac{dx_2}{x_2} = \frac{dx_1}{x_1}$$

so that $x_2 = cx_1$ and the solution lies on a straight line through the origin $x_1 = x_2 = 0$, $(r = 0)$. In the (x_1, x_2) plane the picture we have is that the solution must move along a line making some constant angle α with the x_1-axis; see Figure 7–16. If $\cos\,\alpha \neq 0$, then $r = \sqrt{x_1^2 + x_2^2} = x_1/\cos\,\alpha$;

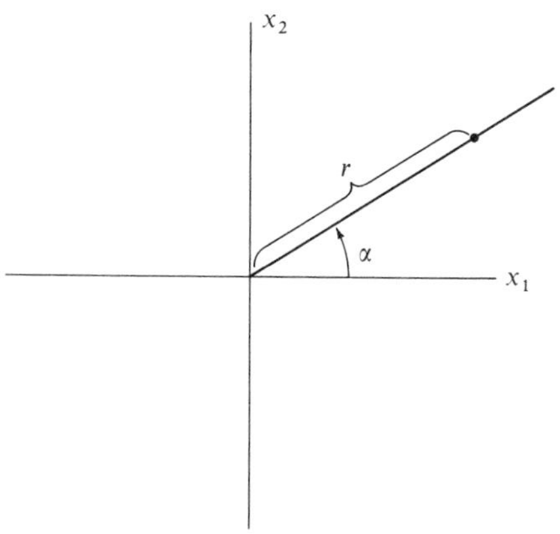

FIGURE 7–16

if $\cos\alpha = 0$, then $r = x_2^2/\sin\alpha$ and $\sin\alpha \neq 0$. In either case, from one of the differential equations of the system we obtain $r'' = -(1/r^2)$. If we begin at $r = r_0 > 0$ with $r' = 0$ at $t = 0$, then it is not hard to see that after some *finite* lapse of time the solution point $r(t; (r_0, 0))$ will "collide" with the singularity.

Write $r_1 = r$, $r_2 = r'$; then

$$\begin{aligned} r_1' &= r_2 \\ r_2' &= -\frac{1}{r_1^2}. \end{aligned} \tag{7–9}$$

We have $r_2 r_2' + (1/r_1^2)\, r_1' = 0$ along solutions and so

$$\tfrac{1}{2}r_2^2 - \frac{1}{r_1} = \text{constant} = c \tag{7–10}$$

is an integral of the system in Eq. 7–9. This shows that as $r_1 = r \to 0$, the velocity $r_2 = r'$ becomes infinite. However, writing Eq. 7–10 as $r_2^2 = 2(c + (1/r_1))$, we have $r' = \pm\sqrt{2(c + (1/r))}$; taking the negative branch for a collision trajectory we have

$$-\frac{dr}{\sqrt{2\left(c + \dfrac{1}{r}\right)}} = dt$$

so that the time elapsed between the motion from the starting point r_0 and the collision at $r = 0$ is

$$t_{r_0} = \int_0^{r_0} \frac{dr}{\sqrt{2\left(c + \dfrac{1}{r}\right)}}. \tag{7–11}$$

The constant c is negative, since from Eq. 7–10 for our initial conditions $r(0) = r_0 > 0$ and $r' = 0$, we have: $r_1 = r_0$, $r_2 = 0$ implies $c = -(1/r_0) < 0$. The integral in Eq. 7–11 can be evaluated explicitly and yields

$$t_{r_0} = \frac{2}{3\sqrt{2}\,c^2}\{2 - (2 + cr_0)\sqrt{1 + cr_0}\};$$

and since $c = -(1/r_0)$, this gives

$$t_{r_0} = \frac{4r_0^2}{3\sqrt{2}}. \tag{7–12}$$

Thus the collision takes place after a finite lapse of time given by Eq. 7–12.

It can be shown [108] that even these collision trajectories can be analytically continued for all t by extending the domain of definition of t from the real line to a three-sheeted *Riemann surface*. The behavior of the collision trajectory near $t = t_{r_0}$ is expressible as

$$r(t) = a\left(t - \frac{4r_0^2}{3\sqrt{2}}\right)^{2/3} + \dots \tag{7–13}$$

If we call $\tau = t - (4r_0^2/3\sqrt{2})$, then from Eq. 7–13 we have

$$r' = \tfrac{2}{3}a\tau^{-1/3} + \dots$$

$$r'' = \tfrac{2}{3}(-\tfrac{1}{3})a\tau^{-4/3} + \dots$$

$$-\frac{1}{r^2} = -\frac{1}{a^2}\tau^{-4/3} + \dots.$$

Therefore $r'' = -(1/r^2)$ implies $-(2/9)a = -(1/a^2)$ or $a = (9/2)^{1/3}$. In fact, with this value of a the expansion whose first term is shown in Eq. 7–13 expresses $r(t)$ for values of t both less than as well as greater than or equal to $t_{r_0} = (4r_0^2/3\sqrt{2})$. For values of $t > t_{r_0}$ the solution moves away from the collision point $r = 0$. Thus the analytic continuation of a collision trajectory produces an ejection trajectory.

EXERCISE

Compare numerical solutions to the two-body problem for a trajectory passing near the singularity at $r = 0$ with and without the use of the transformation $\tau = t - (4r_0^2/3\sqrt{2})$ and Eq. 7–13. Use the equations

$$x_1'' = -\frac{x_1}{r^3}, \qquad r = (r_1^2 + x_2^2)^{1/2}, \qquad x_2'' = -\frac{x_2}{r^3}$$

for this numerical experiment.

8

A PRIORI LOCAL TRANSFORMATIONS

8.1 INTRODUCTION

In the discussion centering on Figure 7–7 in Section 7.5 we derived, in effect, different transformations in each of certain regions in space. In the restricted three-body problem, the Jacobi integral of Eq. 7–7 could be used as a transformation to reduce the order of the system from four to three, taking care to stay on the correct part of the surface when alternating expressions. Unlike the simpler two-body problem with its two integrals, the three-body problem can use only the Jacobi integral—there are no other simple global integrals to use to reduce the order further. It is possible, however, to make use of *local* integrals to perform local transformations which reduce a system of order n to one of order one.

8.2 LOCAL INTEGRALS

Given an autonomous system

$$\mathbf{y}' = \mathbf{f}(\mathbf{y}), \qquad \mathbf{y} \in E^n$$

with the components $f_i(\mathbf{y})$ of $\mathbf{f}(\mathbf{y})$ analytic in the components $y_1, \ldots, y_n$ of $\mathbf{y}$ near the vector $\mathbf{y}(0) = \mathbf{Y}$, and given that $\mathbf{f}(\mathbf{Y}) \neq \mathbf{0}$, then there exist $n - 1$ independent functions $H_1(\mathbf{y}), \ldots, H_{n-1}(\mathbf{y})$, analytic in $\mathbf{y}$ near $\mathbf{Y}$ (that is, expressible in power series), which are called *local integrals*; these functions have the property that for a trajectory $\mathbf{y}(t)$ through $\mathbf{Y}$ we have

$$\frac{d}{dt} H_i(\mathbf{y}) = 0$$

for $\mathbf{y}$ sufficiently near $\mathbf{Y}$.

EXAMPLE Consider the trivial equation

$$y_1' = y_2$$
$$y_2' = -y_1$$

with the initial data

$$\mathbf{Y} = \begin{pmatrix} 0 \\ 1 \end{pmatrix}.$$

The function

$$H(y_1, y_2) = y_2 - (1 - y_1^2)^{1/2}, \qquad |y_1| < 1$$

is a local integral since

$$\frac{d}{dt} H(y_1, y_2) = y_1(1 - y_1^2)^{-1/2}(y_2 - \sqrt{1 - y_1^2})$$

and $y_1^2 + y_2^2 = 1$ and therefore $(d/dt)H(y_1, y_2) = 0$ on the trajectory. The integral is only *local* in this form, since the square root is not defined for $|y_1| > 1$ and since the surface is only an integral for $\mathbf{y}(0) = \mathbf{Y}$. ●

A general procedure is known [20] for computing the coefficients in the series expansions for the local integrals. It is derived in a straightforward way from the defining relation for the local integral substituting series expansions and collecting terms. From the assumption that $\mathbf{Y}$ is not a rest point, it follows that at least one of the components of $\mathbf{f}(\mathbf{Y})$ does not vanish, say $f_1(\mathbf{Y}) \neq 0$. If $y_1 = y_1(t, \mathbf{Y}) = Y_1 + f_1(\mathbf{Y})t + O(t^2)$, then for small t we have approximately

$$t = \frac{y_1 - Y_1}{f_1(\mathbf{Y})}.$$

From this we deduce that $n - 1$ linearly independent planes defined by

$$y_k - Y_k - f_k(\mathbf{Y})\left(\frac{y_1 - Y_1}{f_1(\mathbf{Y})}\right) = 0,$$

for $k = 2, 3, \ldots, n$, have the property that $\mathbf{y}(t, \mathbf{Y})$ remains close to each of them to order $O(\{y_1 - Y_1\}^2)$. The intersection of these planes is the straight line given parametrically by the equations

$$
\begin{aligned}
y_2 &= Y_2 + f_2(\mathbf{Y})\left(\frac{y_1 - Y_1}{f_1(\mathbf{Y})}\right) \\
y_n &= Y_n + f_n(\mathbf{Y})\left(\frac{y_1 - Y_1}{f_1(\mathbf{Y})}\right).
\end{aligned}
\tag{8–1}
$$

Writing this as

$$
\left.\begin{aligned}
y_1 &= Y_1 + f_1(\mathbf{Y})t \\
y_2 &= Y_2 + f_2(\mathbf{Y})t \\
&\cdots\cdots\cdots \\
y_n &= Y_n + f_n(\mathbf{Y})t
\end{aligned}\right\} \qquad \text{or} \qquad \mathbf{y} = \mathbf{Y} + \mathbf{f}(\mathbf{Y})t,
$$

We can see that the straight line given by Eq. 8–1 passes through $\mathbf{Y}$ in the direction $\mathbf{f}(\mathbf{Y})$ given by the vector field at $\mathbf{Y}$. This straight line is tangent to the solution $\mathbf{y}(t, \mathbf{Y})$ at $t = 0$.

This derivation of the local integrals, here written out only through second-order terms, leads to the following expressions.

Assume again that $f_1(\mathbf{Y}) \neq 0$. Define $y_1^*, y_2^*, \ldots, y_n^*$ by

$$
\begin{aligned}
y_1^* &= y_1 - Y_1 \\
y_2^* &= y_2 - Y_2 - \frac{f_2(\mathbf{Y})}{f_1(\mathbf{Y})} y_1^* \\
&\cdots\cdots\cdots\cdots \\
y_n^* &= y_n - Y_n - \frac{f_n(\mathbf{Y})}{f_1(\mathbf{Y})} y_1^*
\end{aligned}
\tag{8–2}
$$

and define $f_1^*(\mathbf{y}^*), \ldots, f_n^*(\mathbf{y}^*)$ by

$$
\begin{aligned}
f_1^*(\mathbf{y}^*) &= f_1(\mathbf{y}) \\
f_2^*(\mathbf{y}^*) &= f_2(\mathbf{y}) - \frac{f_2(\mathbf{Y})}{f_1(\mathbf{Y})} f_1(\mathbf{y}) \\
&\cdots\cdots\cdots\cdots \\
f_n^*(\mathbf{y}^*) &= f_n(\mathbf{y}) - \frac{f_n(\mathbf{Y})}{f_1(\mathbf{Y})} f_1(\mathbf{y}).
\end{aligned}
$$

Define a matrix of elements $a_j^{(i)}$, $(i = 1, 2, \ldots, n, j = 1, 2, \ldots, n)$ by

$$a_j^{(i)} = \left[\frac{\partial f_i^*(\mathbf{y}^*)}{\partial y_j^*}\right]_{\mathbf{y}^*=0}.$$

Define the $n - 1$ quadratic expressions $H_i^{(2)}(\mathbf{y}^*)$, $i = 2, 3, \ldots, n$, by

$$H_i^{(2)}(\mathbf{y}^*) = y_i^* - \frac{1}{f_1(\mathbf{Y})}\left(\tfrac{1}{2}a_1^{(i)}y_1^{*2} + \sum_{j=2}^{n} a_j^{(i)} y_1^* y_j^*\right). \tag{8–3}$$

The surfaces $H_i^{(2)}(\mathbf{y}^*) = 0$ are quadratic approximations to the local integrals in the sense that the solution $\mathbf{y}(t, \mathbf{Y}_0)$ substituted into Eq. 8–2, and Eq. 8–3 leads to

$$H_i^{(2)}(\mathbf{y}^*(t, \mathbf{Y}_0)) = O(\|\mathbf{y}^*\|^3).$$

The planes discussed in a previous paragraph corresponding to the separate equations in Eq. 8–1 can be expressed, using the notation of Eq. 8–2, in terms of the linear mappings

$$H_i^{(1)}(\mathbf{y}^*) \equiv y_i^*,\ i = 2, 3, \ldots, n,$$

as

$$H_i^{(1)}(\mathbf{y}^*) = 0,\ i = 2, 3, \ldots, n.$$

Surfaces of degree k, $H_i^{(k)}(\mathbf{y}^*) = 0$, can be derived by following the same procedure; the expressions $H_i^{(k)}(\mathbf{y}^*)$ converge to local integrals $H_i(\mathbf{y}^*)$ as $k \to \infty$. The meaning of *local* is once again that the expressions $H_i^{(k)}(\mathbf{y}^*)$ are only defined near $\mathbf{y}^* = 0$, that is, near $\mathbf{y} = \mathbf{Y}$, and the convergence $H_i^{(k)}(\mathbf{y}^*) \to H_i(\mathbf{y}^*)$ only holds for small enough $\|\mathbf{y}^*\|$.

EXAMPLE Consider again the system

$$y_1' = y_2$$

$$y_2' = -y_1$$

and the initial point $\mathbf{Y} = (0, 1)$. We obtain $Y_1 = 0$, $Y_2 = 1$, $f_1(\mathbf{Y}) = Y_2 = 1$, $f_2(\mathbf{Y}) = 0$ and so

$$y_1^* = y_1$$

$$y_2^* = y_2 - 1$$

$$f_1^*(\mathbf{y}^*) = y_2^* + 1$$

$$
\begin{aligned}
&f_2^*(\mathbf{y}^*) = -y_1^* \\
&a_1^{(1)} = 0, \qquad a_2^{(1)} = 0 \\
&a_1^{(2)} = -1, \qquad a_2^{(2)} = 0 \\
&H_2^{(2)}(\mathbf{y}^*) = y_2^* + \tfrac{1}{2}y_1^{*2} = y_2 - 1 + \tfrac{1}{2}y_1^2.
\end{aligned}
\tag{8–4}
$$

We recognize the resulting expression for $H_2^{(2)}$ as the first few terms in a power-series expansion of the local integral $H(y_1, y_2) = y_2 - \sqrt{1 - y_1^2}$, $|y_1| < 1$, found earlier at the start of this section. We can define a *local transformation of variables* for this problem using Eq. 8–4 in the following way. For small $|y_1|$, $H_2^{(2)}(\mathbf{y}^*)$ given in Eq. 8–4 is *nearly* constant along the solution $\mathbf{y}(t, \mathbf{y}_0)$. We therefore set

$$
\begin{aligned}
z_1 &= y_1 \\
z_2 &= H_2^{(2)}(\mathbf{y}^*) = y_2 - 1 + \tfrac{1}{2}y_1^2
\end{aligned}
\tag{8–5}
$$

and derive the differential equations

$$
\begin{aligned}
z_1' &= y_1' = y_2 = z_2 + 1 - \tfrac{1}{2}z_1^2 \\
z_2' &= y_2' + y_1 y_1' = -z_1 + z_1(z_2 + 1 - \tfrac{1}{2}z_1^2)
\end{aligned}
$$

or

$$
\begin{aligned}
z_1' &= 1 - \tfrac{1}{2}z_1^2 + z_2 \\
z_2' &= z_1(z_2 - \tfrac{1}{2}z_1^2).
\end{aligned}
\tag{8–6}
$$

The initial point corresponding to $\mathbf{Y} = (0, 1)$ is $\mathbf{z}(0) = (0, 0)$. It is not clear that anything of interest is gained by this rather complicated transformation process for the simple equations of this example.

EXERCISE

Try a numerical experiment and see!

Indeed we have gone from the *linear* system $y_1' = y_2$, $y_2' = -y_1$ to the *nonlinear* system in Eq. 8–6. We chose such a simple example, however, to illustrate the mechanism of the expression in Eq. 8–3. In order to restore a little coin in the idea that these local integrals may have some use, we point out that from Eq. 8–6 we have

$$z_1'(0) = 1, \qquad z_2'(0) = 0;$$

and if we approximate the solution of Eq. 8–6 to terms of first order in t, $z_1 \equiv t$, $z_2 \equiv 0$ we obtain, from Eq. 8–5, the approximate solution, in the original variables, $\bar{y}_1 = t$, $\bar{y}_2 = 1 - (1/2)t^2$ to be compared with the exact solution $y_1(t, y_0) = \sin t$, $y_2(t, y_0) = \cos t$. We have

$$\bar{y}_1 - y_1(t, Y) = O(t^3)$$

$$\bar{y}_2 - y_2(t, Y) = O(t^4). \quad \bullet$$

More generally, a local transformation defined by $z_1 = y_1$ and, for $i = 2, 3, \ldots, n$, by $z_i = H_i^{(2)}(\mathbf{y}^*)$, defined by the expression in Eq. 8–3, leads to a set of differential equations $z_i' = g_i(\mathbf{z})$ derived from a given set $y_i' = f_i(\mathbf{y})$ with $f_1(\mathbf{Y}) \neq 0$ as follows. From Eq. 8–3 we obtain for $i = 2, 3, \ldots, n$

$$
\begin{aligned}
z_i' = g_i(\mathbf{z}) &= \{H_i^{(2)}(\mathbf{y})\}' \\
&= y_i^{*\prime} - \frac{1}{f_1(\mathbf{Y})}\left\{a_1^{(i)} y_1^* y_1^{*\prime} + \sum_{j=2}^{n} a_j^{(i)} (y_1^* y_j^{*\prime} + y_1^{*\prime} y_j^*)\right\}
\end{aligned}
\tag{8–7}
$$

and

$$z_1' = g_1(\mathbf{z}) = y_1' = f_1(\mathbf{y}).$$

From the defining transformation $z_i = H_i^{(2)}(\mathbf{y}^*)$, $i = 2, \ldots, n$ and the Eq. 8–3 we write out the system

$$\left(1 - \frac{a_2^{(2)} y_1^*}{f_1(\mathbf{Y})}\right) y_2^* - \frac{a_3^{(2)} y_1^*}{f_1(\mathbf{Y})} y_3^* \ldots - \frac{a_n^{(2)} y_1^*}{f_1(\mathbf{Y})} y_n^* = z_2 + \frac{a_1^{(2)} y_1^{*2}}{2f_1(\mathbf{Y})}$$

$$\cdots\cdots\cdots\cdots\cdots\cdots\cdots\cdots\cdots\cdots$$

$$\frac{-a_2^{(n)} y_1^*}{f_1(\mathbf{Y})} y_2^* - \frac{a_3^{(n)} y_1^*}{f_1(\mathbf{Y})} y_3^* \ldots + \left(1 - \frac{a_n^{(n)} y_1^*}{f_1(\mathbf{Y})}\right) y_n^* = z_n + \frac{a_1^{(n)} y_1^{*2}}{2f_1(\mathbf{Y})}$$

The matrix of coefficients is nonsingular for small y_1^*, and the system can be solved uniquely for $y_2^*, y_3^*, \ldots, y_n^*$. Using Eq. 8–2 we can then find $y_1, y_2, \ldots, y_n$ as functions of $z_1, z_2, \ldots, z_n$. Recall that $y_1 = z_1$. The resulting expressions can be then substituted into Eq. 8–7 to obtain expressions for $g_i(\mathbf{z})$. If we approximate $g_i(\mathbf{z})$ by $g_i(0)$, we will obtain, just as in the simple example discussed, an approximate solution $\bar{\mathbf{y}}$, after transforming back to original variables, such that $\|\bar{\mathbf{y}} - \mathbf{y}(t, \mathbf{y}_0)\| = O(t^3)$. There are undoubtedly many other ways to use the local integrals or approximation to them. The derivations indicated just now for the expressions $g_i(\mathbf{z})$ could be programmed for the computer in either symbolic or numerical form. The importance of the technique lies in the reduction to a first-order equation for one unknown and in the high accuracy in $\mathbf{y}$ obtained from lower accuracy in $\mathbf{z}$. This approach is neither well known nor widely used and still awaits numerical experimentation for evidence of its possible practical value.

EXERCISE

Derive the transformation explicitly for the nonlinear pendulum,

$$y_1' = y_2$$

$$y_2' = -\sin y,$$

and devise a numerical experiment to test its possible value.

8.3 THE CIRCLE-OF-CURVATURE TRANSFORMATION

Consider again the system

$$\mathbf{y}' = \mathbf{f}(\mathbf{y}), \qquad \mathbf{y} \in E^n, \qquad \mathbf{y}(0) = \mathbf{Y}$$

defining a vector field and flow in E^n and a trajectory through the point $\mathbf{Y}$. We wish to find a transformation which reduces the gross primary rotation of the flow near $\mathbf{Y}$. Suppose that $\mathbf{y}'$ and $\mathbf{y}''$ are not collinear at $t = 0$; for convenience we write these *vectors* as $\mathbf{y}_0'$ and $\mathbf{y}_0''$, not to be confused with y_i', $i \geq 1$, a *component* of $\mathbf{y}'$. The vectors $\mathbf{y}_0'$ and $\mathbf{y}_0''$ determine the osculating plane to the curve $\mathbf{y}(t)$ through $\mathbf{y}_0 \equiv \mathbf{Y} = \mathbf{y}(0)$. Writing $\langle \mathbf{u}, \mathbf{v} \rangle$ for the real-valued inner product of two vectors $\mathbf{u}$ and $\mathbf{v}$ in E^n, $\langle \mathbf{u}, \mathbf{v} \rangle = u_1 v_1 + \ldots + u_n v_n$, we represent the *circle of curvature* to $\mathbf{y}(t)$ at $\mathbf{y}_0 = \mathbf{Y}$ as a circle in the $(\mathbf{y}_0', \mathbf{y}_0'')$ plane having its center at

$$\mathbf{c}_0 = \mathbf{y}_0 + \rho_0 \frac{\mathbf{s}_0}{\langle \mathbf{s}_0, \mathbf{s}_0 \rangle^{1/2}}$$

and its radius ρ_0 equal to

$$\rho_0 = \frac{\langle \mathbf{y}_0', \mathbf{y}_0' \rangle^{1/2}}{\langle \mathbf{s}_0, \mathbf{s}_0 \rangle^{1/2}},$$

where

$$\mathbf{s}_0 = \mathbf{y}'' - \frac{\langle \mathbf{y}_0', \mathbf{y}_0'' \rangle}{\langle \mathbf{y}_0', \mathbf{y}_0' \rangle} \mathbf{y}_0'$$

is a vector normal to the circle of curvature. The unit-tangent vector $\boldsymbol{\tau}(0)$ and unit-normal vector $\boldsymbol{\nu}(0)$ at $\mathbf{y}_0 = \mathbf{Y}$ are given by

$$\boldsymbol{\tau}(0) = \frac{\mathbf{y}_0'}{\langle \mathbf{y}_0', \mathbf{y}_0' \rangle^{1/2}}$$

$$\boldsymbol{\nu}(0) = \frac{\mathbf{s}_0}{\langle \mathbf{s}_0, \mathbf{s}_0 \rangle^{1/2}}.$$

We wish now to find a transformation to make the calculation of the trajectory through $\mathbf{y}_0 = \mathbf{Y}$ easier. Because of computational errors, lack of precise numbers in the data for many physical problems, and the desire to see how errors propagate, it is of interest to consider trajectories through points $\mathbf{W}$ near $\mathbf{Y}$ as well as the trajectory through $\mathbf{Y}$ itself. From this viewpoint we are considering the equations as defining a flow, and we wish to see how small packets of the "fluid" move, not just individual points. Thus we consider the trajectory $\mathbf{y}(t; \mathbf{W})$ satisfying $\mathbf{y}(0; \mathbf{W}) = \mathbf{W}$, where $\mathbf{W}$ is near $\mathbf{Y}$. We approximate this near $t = 0$ by a rotational motion $\mathbf{y}^*$ with constant angular velocity in the $(\mathbf{y}_0', \mathbf{y}_0'')$ plane about the center point $\mathbf{c}_0$.

We put

$$\mathbf{y}^*(t; \mathbf{W}) = \mathbf{c}_0 + \langle \mathbf{W} - \mathbf{c}_0, \boldsymbol{\tau}(0) \rangle \boldsymbol{\tau}(t) + \langle \mathbf{W} - \mathbf{c}_0, \boldsymbol{\nu}(0) \rangle \boldsymbol{\nu}(t) + \mathbf{e} \qquad (8\text{–}8)$$

where

$$\boldsymbol{\tau}(t) = \boldsymbol{\tau}(0) \cos \theta(t) + \boldsymbol{\nu}(0) \sin \theta(t)$$

$$\boldsymbol{\nu}(t) = -\boldsymbol{\tau}(0) \sin \theta(t) + \boldsymbol{\nu}(0) \cos \theta(t),$$

$$\mathbf{e} = \mathbf{W} - \mathbf{c}_0 - \langle \mathbf{W} - \mathbf{c}_0, \boldsymbol{\tau}(0) \rangle \boldsymbol{\tau}(0) - \langle \mathbf{W} - \mathbf{c}_0, \boldsymbol{\nu}(0) \rangle \boldsymbol{\nu}(0),$$

$$\theta(t) = \frac{\langle \mathbf{y}_0', \mathbf{y}_0' \rangle^{1/2}}{\rho_0} t.$$

We choose a $\mathbf{z}$ coordinate system to be carried along by the approximating rotational flow $\mathbf{y}^*$ and represent $\mathbf{y}(t; \mathbf{W})$ by

$$\mathbf{y}(t; \mathbf{W}) = \mathbf{c}_0 + z_1 \boldsymbol{\tau}(t) + z_2 \boldsymbol{\nu}(t) + \hat{\mathbf{z}}$$

with

$$z_1(0) = \langle \mathbf{W} - \mathbf{c}_0, \boldsymbol{\tau}(0)\rangle \text{ in } E^1,$$

$$z_2(0) = \langle \mathbf{W} - \mathbf{c}_0, \boldsymbol{\nu}(0)\rangle \text{ in } E^1,$$

$$\hat{\mathbf{z}}(0) = \mathbf{e} \text{ in } E^n, \qquad \langle \hat{\mathbf{z}}(t), \boldsymbol{\tau}(t)\rangle = \langle \hat{\mathbf{z}}(t), \boldsymbol{\nu}(t)\rangle = 0.$$

Notice that for $n = 2$, $\hat{\mathbf{z}}(t) = 0$ since $\boldsymbol{\tau}(t)$, $\boldsymbol{\nu}(t)$ form a basis for E^2. Using this transformation we get an equation in the $\mathbf{z}$ coordinates,

$$z_1' = \theta_0' z_2 + \langle \hat{\mathbf{z}}(t), \boldsymbol{\nu}(t)\rangle\theta_0' + \langle \mathbf{f}(\mathbf{c}_0 + z_1(t)\boldsymbol{\tau}(t) + z_2(t)\boldsymbol{\nu}(t) + \hat{\mathbf{z}}(t)), \boldsymbol{\tau}(t)\rangle,$$

$$z_2' = -\theta_0' z_1 - \langle \hat{\mathbf{z}}(t), \boldsymbol{\tau}(t)\rangle\theta_0' + \langle \mathbf{f}(\mathbf{c}_0 + z_1(t)\boldsymbol{\tau}(t) + z_2(t)\boldsymbol{\nu}(t) + \hat{\mathbf{z}}(t)), \boldsymbol{\nu}(t)\rangle,$$

$$\hat{\mathbf{z}}' = \mathbf{f}(\mathbf{c}_0 + z_1(t)\boldsymbol{\tau}(t) + z_2(t)\boldsymbol{\nu}(t) + \hat{\mathbf{z}}(t)) + (\theta_0' z_2 - z_1')\boldsymbol{\tau}(t) - (\theta_0' z_1 + z_2')\boldsymbol{\nu}(t).$$

The first of these two are scalar equations, while the third is an nth order system for $\hat{\mathbf{z}}$, which is known to lie in the $n - 2$ dimensional space orthogonal to $\boldsymbol{\tau}(t)$ and $\boldsymbol{\nu}(t)$. By θ_0' in the above we mean

$$\theta_0' = \frac{\langle \mathbf{y}_0', \mathbf{y}_0'\rangle^{1/2}}{\rho_0} = \theta'(0).$$

If the transformation of Eq. 8–8 gives a good approximation to the flow near $t = 0$, $\mathbf{y} = \mathbf{Y}$, then the $\mathbf{z}$-flow will have very little rotation in it and computationally will be simple to handle. A solution to the original initial value problem can be computed by setting $\mathbf{W}$ equal to $\mathbf{Y}$.

EXAMPLE As a very simple illustrative example, consider again the system

$$y_1' = y_2$$

$$y_2' = -y_1$$

with $\mathbf{Y} = (0, 1)$. We find then that $\mathbf{y}_0' = (1, 0)$, $\mathbf{y}_0'' = (0, -1)$, $\langle \mathbf{y}_0', \mathbf{y}_0''\rangle = 0$, $\mathbf{s}_0 = (0, -1)$, $\rho_0 = 1$, $\boldsymbol{\tau}(0) = (1, 0)$, $\boldsymbol{\nu}(0) = (0, -1)$, $e = 0$, $\mathbf{c}_0 = (0, 0)$. Therefore we have $\theta(t) = t$, $\theta_0' = 1$, $\boldsymbol{\tau}(t) = (\cos t, -\sin t)$, and $\boldsymbol{\nu}(t) = (-\sin t, -\cos t)$. Under our transformation

$$\mathbf{y}(t; \mathbf{W}) = z_1\boldsymbol{\tau}(t) + z_2\boldsymbol{\nu}(t) + \hat{\mathbf{z}}$$

we find the differential equations for $\mathbf{z}$ to be

$$z_1' = 0$$

$$z_2' = 0;$$

$\hat{\mathbf{z}}(t) \equiv 0$ since we are in E^2, that is, $n = 2$. Thus

$$z_1(t) = W_1$$

$$z_2(t) = -W_2$$

for all t; this transformation has eliminated *all* rotation from the flow. For more complex systems, of course, rotation and expansion will only be reduced. ●

EXERCISE

Derive the circle-of-curvature transformations for the nonlinear-pendulum equations.

EXERCISE

Devise and carry out a numerical experiment to test the possible value of the transformation on the nonlinear-pendulum equations.

Additional reference: 80.

9

A PRIORI TIME-DEPENDENT TRANSFORMATIONS

9.1 INTRODUCTION

We have already mentioned time-dependent transformations of the simplest kind, namely those based in essence on perturbation ideas. Thus, to solve

$$\mathbf{y}' = \mathbf{a} + \mathbf{g}(\mathbf{y}), \qquad \mathbf{y} \in E^n,$$

we might let

$$\mathbf{y} = \mathbf{z} + \mathbf{a}t$$

and solve for $\mathbf{z}$ from

$$\mathbf{z}' = \mathbf{g}(\mathbf{z} + \mathbf{a}t).$$

We wish in this chapter, however, to consider more complicated types of transformations.

9.2 TRANSFORMING *t*

We shall consider very briefly the possible benefit in a simple transformation of the independent variable t to a new independent variable τ. Suppose we have a transformation defined implicitly by

$$t = s(\tau).$$

Given a differential equation

$$\frac{d\mathbf{y}}{dt} = \mathbf{f}(t, \mathbf{y}),$$

we define a new unknown $\mathbf{z}(\tau)$ by

$$\mathbf{z}(\tau) = \mathbf{y}(t) = \mathbf{y}(s(\tau)).$$

Then we derive a differential equation for $\mathbf{z}(\tau)$, namely

$$\frac{d\mathbf{z}}{d\tau} = \frac{ds}{d\tau}\,\mathbf{f}(s(\tau), \mathbf{z}(\tau)).$$

By a suitable choice of the transformation $t = s(\tau)$, the equation for $\mathbf{z}(\tau)$ may be easier to solve than that for $\mathbf{y}(t)$.

EXERCISE

Consider the differential equation $y' = y^2 + yt^{-1/2}$. Find a transformation $t = s(\tau)$ which removes the singularity at $t = 0$ for any $y(0)$ by examining the dependence of y on t near $t = 0$.

If a finite-difference method using equal steps h is used in the variable τ, this leads to generally nonconstant steps

$$t_{i+1} - t_i = s((i + 1)h) - s(ih)$$

in the variable t. In fact, the numerical methods for varying the step size during execution of a method can be viewed as a run-time step-by-step definition of the transformation $t = s(\tau)$. The usual practice is to choose the transformation (equivalently, the variable step size) to keep some estimate of local truncation error within certain tolerances. Consider the use of a variable-step Taylor-series method. Here we use Taylor expansions through the kth power at points $0 = t_0 < t_1 < \ldots$ to compute vectors $\mathbf{Y}_i$ approximating the solution $\mathbf{y}(t_i)$ via

$$\mathbf{Y}_{i+1} = \sum_{j=0}^{k} \frac{(t_{i+1} - t_i)^j}{j!}\,\mathbf{y}_i^{(j)},$$

where

$$\mathbf{y}_i^{(j)} = \left[\frac{d^j}{(dt)^j}\,\mathbf{y}(t)\right]_{t = t_i,\, \mathbf{y} = \mathbf{Y}_i}.$$

The step $t_{i+1} - t_i$ is selected so that

$$\frac{\|\mathbf{y}_{i+1}^{(k+1)}\|}{(k+1)!}(t_{i+1} - t_i)^{k+1} = \frac{\|\mathbf{y}_1^{(k+1)}\|}{(k+1)!}\, t_1^{k+1} \leq \epsilon$$

where ϵ is a given tolerance for estimated local error. If the transformation function $s(\tau)$ satisfies

$$s'(\tau_i) \sim \left(\frac{\|\mathbf{y}_1^{(k+1)}\|}{\|\mathbf{y}_i^{(k+1)}\|}\right)^{\frac{1}{k+1}} \frac{t_1}{\tau_1},$$

where $\tau_{i+1} - \tau_i = h$ is constant for all i, then equal spacing in τ would lead to the desired spacing in t. In fact, of course, the easiest way to compute the step in t is via

$$t_{i+1} = t_i + \left(\frac{\|\mathbf{y}_1^{(k+1)}\|}{\|\mathbf{y}_i^{(k+1)}\|}\right)^{\frac{1}{k+1}} t_1,$$

which avoids dealing with the transformation $s(\tau)$; it should be noted that the above formula for t_{i+1} is nothing other than Euler's method for solving the differential equation for $s(\tau)$. Other methods might give better results.

Using such procedures for the automatic variation of step size in numerical solutions of the initial value problem on computers can result in spectacular improvement in efficiency. In numerical solutions of space-probe trajectory problems, for instance, it is often possible to increase the step size used near the starting point (say near the earth) by a thousandfold or more as the space probe moves far away from the earth's gravitational center (which is a singularity of the vector field). Eventually, as the probe nears another planet; the step size would be decreased again according to the relation given in this section, since $y_p^{(k+1)}$ would begin increasing again. In order to obtain results of comparable accuracy with the use of the same algorithm with a *fixed* step length in t, it might take a hundred or even a thousand times as much computing time.

EXERCISE

Try it!

There are a number of schemes in use which involve occasional halving or doubling of step size according to various criteria. We have not found these as effective as the smoother variations given by formulas such as the one in this section.

Additional reference: 80.

9.3 TRANSFORMATIONS BY SERIES

In Sections 4.3 and 5.8 we indicated that, by using formal differentiations of the right-hand side, one could develop Taylor-series numerical methods of high accuracy. It is also possible to combine Taylor series with other numerical algorithms in order to create a still more accurate method. Consider the time-dependent system

$$\mathbf{y}' = \mathbf{f}(t, \mathbf{y}), \qquad \mathbf{y} \in E^n.$$

Suppose we have the solution, or an approximate one, at time $t = t_0$ with $\mathbf{y}(t_0) = \mathbf{y}_0$. Let $\mathbf{y}_0^{(1)}, \mathbf{y}_0^{(2)}, \ldots, \mathbf{y}_0^{(m+1)}$ denote the first $m + 1$ derivatives of $\mathbf{y}(t)$ at $t = t_0$ evaluated by formal differentiation. By means of a truncated Taylor series we define a transformed dependent variable

$$\mathbf{z}(t) = \mathbf{y}(t) - \sum_{i=0}^{m+1} \frac{\mathbf{y}_0^{(i)}}{i!} (t - t_0)^i.$$

Then $\mathbf{z}(t)$ satisfies the differential equation

$$\mathbf{z}' = \mathbf{g}(t, \mathbf{z}) \equiv \mathbf{f}\left(t, \mathbf{z} + \sum_{i=0}^{m+1} \frac{\mathbf{y}_0^{(i)}}{i!} (t - t_0)^i\right) - \sum_{i=0}^{m+1} \frac{\mathbf{y}_0^{(i)}}{(i-1)!} (t - t_0)^{i-1}$$

and

$$\mathbf{z}_0 = \mathbf{z}_0^{(1)} = \mathbf{z}_0^{(2)} = \ldots = \mathbf{z}_0^{(m+1)} = 0.$$

Because of the vanishing of the $m + 1$ derivatives of $\mathbf{z}(t)$ at $t = t_0$, we may be able to compute $\mathbf{z}$ more accurately by any numerical method, at least for the $t - t_0$ not too large.

EXERCISE

For an illustration, try this with Euler's method on $y' = y^2$, $y(0) = 1$. Compare numerical results with and without the transformation for several values of m.

One type of numerical method which has been applied by E. Fehlberg

[41, 42] to the transformed equation for $\mathbf{z}$ with great success is the Runge-Kutta family of methods. Given a step size h, if we let

$$\mathbf{k}_1 = h\mathbf{g}(t_0 + a_1 h, \mathbf{z}_0)$$

$$\mathbf{k}_2 = h\mathbf{g}(t_0 + a_2 h, \mathbf{z}_0 + b_1\mathbf{k}_1)$$

$$\mathbf{k}_3 = h\mathbf{g}(t_0 + a_3 h, \mathbf{z}_0 + b_2\mathbf{k}_1 + b_3\mathbf{k}_2),$$

where

$$a_1 = 1, \qquad a_2 = \frac{m+2}{m+4}, \qquad a_3 = 1,$$

$$b_1 = \frac{1}{m+4}\left(\frac{m+2}{m+4}\right)^{m+1}, \qquad b_2 = \frac{-1}{m+2}, \qquad b_3 = \frac{2}{m+2}\left(\frac{m+4}{m+2}\right)^{m+1},$$

then setting

$$c_1 = \tfrac{1}{2}\frac{m+4}{(m+2)(m+3)}\left(\frac{m+4}{m+2}\right)^{m+1}, \qquad c_2 = \tfrac{1}{2}\frac{1}{m+3}$$

yields

$$\mathbf{z}(t_0 + h) = \mathbf{z}(t_0) + c_1\mathbf{k}_2 + c_2\mathbf{k}_3 + O(h^{m+5}).$$

That is, we get a method with global error of order $m + 4$ using three evaluations of the right-hand-side $\mathbf{g}$ per step; the order can be increased to $m + 5$ with two further evaluations per step. Numerical experiments using these methods have indicated that they require less time, including transformation time, than standard methods such as regular Taylor series or Runge-Kutta to achieve comparable accuracy. Step-size control can be implemented easily. For example [39, 46], on the problem of the heavy asymmetrical top described by a system of six equations, a twelfth-order method of this type required less computing time than any of several other methods, including Runge-Kutta formulas due to Huta or Shanks and Taylor-series methods of several different orders. Of the methods compared, the one described above required less computing time than the other methods of the same order for orders of six, eight, and twelve.

EXERCISE

Devise and carry out a numerical experiment using the family of methods just described for various values of m. Find a value for m by experimentation which gives minimum computing time to achieve a given accuracy on some problem, say to find $y(.5)$ where $y' = y^2$, $y(0) = 1$.

10

A POSTERIORI TRANSFORMATIONS

10.1 INTRODUCTION

In the three preceding chapters we have considered techniques for transforming a problem so that we can solve that particular problem more efficiently. In very many practical situations, however, one really wants to solve a large number of "nearby" problems as well; in design problems, for example, one wants to examine and study the nature of solutions when various parameters are varied by (generally) small amounts. For such problems it is often possible to make use of one tediously computed solution in such a way as to reduce the difficulty involved in computing the nearby solutions to the equations with slightly different parameters. We call transformations designed to do this *a posteriori transformations* since they occur only after some computation.

For simplicity we shall restrict ourselves to the study of a posteriori transformations designed to simplify problems in which only the initial data is varied, not the equation itself; the methods apply with some modification in detail—but not in spirit—to problems in which the equations themselves vary.

In Section 5.8 we discussed error bounding by means of the techniques of interval analysis. The basic idea there was to bound the solution inside n-dimensional rectangles with faces parallel to the coordinate planes; this has the inherent difficulty of causing growing rectangles when trajectories rotate. For example, consider the family of trajectories which are circles about the origin in E^2. A small square of side ϵ about the point (0, 1) would rotate around to $(\sqrt{2}/2, \sqrt{2}/2)$, at which time the smallest bounding square with sides parallel to the coordinate axes has sides of length $\epsilon\sqrt{2}$; in n dimensions the growth would be to $\epsilon\sqrt{n}$. See Figure 10–1.

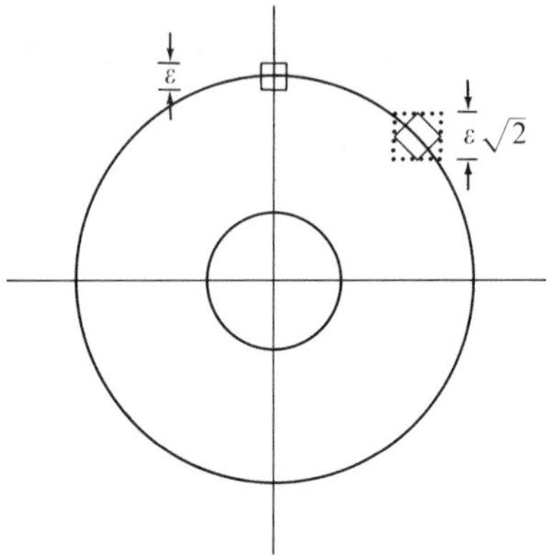

FIGURE 10–1

Therefore, for the purpose of error bounding by interval methods, it is good to reduce the rotation in a system. The transformations to be discussed in this chapter can also be viewed as methods to accomplish this reduction after an approximate solution has been computed.

10.2 INTRINSIC COORDINATES

Consider the system

$$\mathbf{y}' = \mathbf{f}(t, \mathbf{y}), \qquad \mathbf{y} \in E^n;$$

we seek a solution $\mathbf{y}(t; \mathbf{W})$ with $\mathbf{y}(0; \mathbf{W}) = \mathbf{W}$ where $\mathbf{W}$ is near $\mathbf{Y}$ and we know the solution $\mathbf{y}(t; \mathbf{Y})$ with $\mathbf{y}(0; \mathbf{Y}) = \mathbf{Y}$. Regarding $\mathbf{y}(t; \mathbf{Y})$ as a curve in E^n, we can define for this curve a unit tangent vector, a unit normal, a unit binormal, et cetera. More precisely, at $\mathbf{y}(t; \mathbf{Y})$ we can define n unit vectors $\mathbf{e}_1(t), \mathbf{e}_2(t), \ldots, \mathbf{e}_n(t) \in E^n$ by orthonormalizing the vectors $\mathbf{y}', \mathbf{y}''$, $\ldots, \mathbf{y}^{(n)} \in E^n$. These orthonormal vectors, considered as columns, define a matrix $\mathbf{E}(t)$ representing an orthogonal transformation. The flow near $\mathbf{y}(t; \mathbf{Y})$ is then approximated by the flow $\mathbf{y}^*(t; \mathbf{W})$ defined by

$$\mathbf{y}^*(t; \mathbf{W}) = \mathbf{y}(t; \mathbf{Y}) + \mathbf{E}(t)\mathbf{E}(0)^{-1}(\mathbf{W} - \mathbf{Y}).$$

EXERCISE

Derive explicitly as functions of t the components of the matrix $\mathbf{E}(t)$ for $n = 2$.

We define therefore a transformation to the moving coordinates $\mathbf{z}$ by

$$\mathbf{y}(t; \mathbf{W}) = \mathbf{y}(t; \mathbf{Y}) + \mathbf{E}(t)\mathbf{E}(0)^{-1}\mathbf{z}(t; \mathbf{Y}, \mathbf{W}).$$

For $t = 0$ we have $\mathbf{z}(0; \mathbf{Y}, \mathbf{W}) = \mathbf{W} - \mathbf{Y}$ and $\mathbf{z}$ satisfies

$$\mathbf{z}' = \mathbf{E}(0)\mathbf{E}(t)^{-1}\{\mathbf{f}(\mathbf{y}(t; \mathbf{Y}) + \mathbf{E}(t)\mathbf{E}(0)^{-1}\mathbf{z}) - \mathbf{f}(\mathbf{y}(t; \mathbf{Y}))\}$$
$$- \mathbf{E}(0)\mathbf{E}(t)^{-1}\mathbf{E}'(t)\mathbf{E}(0)^{-1}\mathbf{z}.$$

EXERCISE

Verify that if $\mathbf{z}$ satisfies the differential equation shown, then $\mathbf{y}' = \mathbf{f}(t, \mathbf{y})$.

This type of transformation could be used as a perturbation method for computing solutions through $\mathbf{W}$ from a solution through $\mathbf{Y}$. To use it, however, requires first an approximation to $\mathbf{E}(t)$, a difficult problem in itself; it would be possible to approximate $\mathbf{E}(t)$, for small t, by a $\mathbf{E}(t)$ derived from a Taylor-series expansion of $\mathbf{y}(t; \mathbf{Y})$ about $t = 0$.

Additional reference: 80.

10.3 LINEARIZED COORDINATES

We consider approximating the system

$$\mathbf{y}' = \mathbf{f}(t, \mathbf{y}), \qquad \mathbf{y} \in E^n,$$

by the linearized system

$$\mathbf{v}' = \mathbf{A}\mathbf{v} + \mathbf{b}$$

by means of the following procedure. At $t = 0$, $\mathbf{y} = \mathbf{y}_0$ we can evaluate the derivatives

$$\mathbf{y}_0^{(j)} \equiv \left[\frac{d^j}{dt^j}\mathbf{y}(t)\right]_{t=0,\,\mathbf{y}=\mathbf{y}_0}, \qquad j = 1, 2, \ldots, n + 1.$$

We choose the constant matrix $\mathbf{A}$ so that

$$\mathbf{y}_0^{(2)} = \mathbf{A}\mathbf{y}_0^{(1)}$$
$$\vdots$$
$$\mathbf{y}_0^{(n+1)} = \mathbf{A}\mathbf{y}_0^{(n)}$$

and then choose $\mathbf{b}$ so that

$$\mathbf{y}_0^{(1)} = \mathbf{A}\mathbf{y}_0 + \mathbf{b}.$$

If we let $\mathbf{B}$ and $\mathbf{B}'$ be the matrices with columns $\mathbf{y}_0^{(1)}, \ldots, \mathbf{y}_0^{(n)}$ and $\mathbf{y}_0^{(2)}, \ldots, \mathbf{y}_0^{(n+1)}$ respectively, then

$$\mathbf{A} = \mathbf{B}'\mathbf{B}^{-1}, \qquad \mathbf{b} = \mathbf{y}_0^{(1)} - \mathbf{B}'\mathbf{B}^{-1}\mathbf{y}_0.$$

For this choice of $\mathbf{A}$ and $\mathbf{b}$, the solution $\mathbf{v}(t)$ to

$$\mathbf{v}' = \mathbf{A}\mathbf{v} + \mathbf{b}, \qquad \mathbf{v}(0) = \mathbf{y}(0) = \mathbf{Y}$$

will agree in its first $n + 1$ terms with the Taylor series for $\mathbf{y}(t; \mathbf{Y})$.

EXERCISE

Verify the previous statement.

We then consider the matrices $\mathbf{V}(t)$ and $\mathbf{V}'(t)$, whose columns are $\mathbf{v}^{(1)}(t), \mathbf{v}^{(2)}(t), \ldots, \mathbf{v}^{(n)}(t)$ and $\mathbf{v}^{(2)}(t), \mathbf{v}^{(3)}(t), \ldots, \mathbf{v}^{(n+1)}(t)$, respectively, and define a transformation to linearized coordinates

$$\mathbf{y}(t; \mathbf{W}) = \mathbf{y}(t; \mathbf{Y}) + \mathbf{V}(t)\mathbf{V}(0)^{-1}\mathbf{z}(t; \mathbf{Y}, \mathbf{W}).$$

The differential equation for the transformed function $\mathbf{z}(t; \mathbf{Y}, \mathbf{W})$ is now

$$\mathbf{z}' = \mathbf{V}(0)\mathbf{V}(t)^{-1}\{\mathbf{f}(t; \mathbf{Y} + \mathbf{V}(t)\mathbf{V}(0)^{-1}\mathbf{z}) - \mathbf{f}(t; \mathbf{Y})\} - \mathbf{V}(0)\mathbf{V}(t)^{-1}\mathbf{V}'(t)\mathbf{V}(0)^{-1}\mathbf{z}.$$

EXERCISE

Derive the above differential equation for $\mathbf{z}$.

To implement this method we could use an *approximate* solution $\mathbf{y}^*(t; \mathbf{Y})$ rather than $\mathbf{y}(t; \mathbf{Y})$, say a Taylor-series expansion, to define the transformation; the transformed equation for $\mathbf{z}$ remains the same except that $\mathbf{y}^*(t; \mathbf{Y})'$ replaces $\mathbf{f}(t; \mathbf{Y})$.

Additional reference: 80.

10.4 CONNECTION COORDINATES

We are considering in this chapter how the solution changes under changed initial data; we saw in Section 2.5 that this dependence on initial data was described, for small changes $\mathbf{W} - \mathbf{Y}$, by the connection matrix

$$\mathbf{C}(t; \mathbf{Y}) = \frac{\partial}{\partial \mathbf{Y}} \mathbf{y}(t; \mathbf{Y}).$$

For simplicity in the derivations let us suppose now that $\mathbf{f}(t, \mathbf{y}) = \mathbf{f}(\mathbf{y})$ is time-independent. Then $\mathbf{C}(t; \mathbf{Y}) \equiv \mathbf{C}(t)$ satisfies the linear differential equation

$$\mathbf{C}'(t) = \left[\frac{\partial \mathbf{f}}{\partial \mathbf{y}}\right]_{\mathbf{y} = \mathbf{y}(t; \mathbf{Y})} \cdot \mathbf{C}(t),$$

$$\mathbf{C}(0) = \mathbf{I}.$$

For convenience let

$$\mathbf{J}(t) \equiv \mathbf{J}(\mathbf{y}(t; \mathbf{Y})) = \left[\frac{\partial \mathbf{f}}{\partial \mathbf{y}}\right]_{\mathbf{y} = \mathbf{y}(t; \mathbf{Y})}.$$

Then the equation for $\mathbf{C}(t)$ becomes

$$\mathbf{C}'(t) = \mathbf{J}(t)\mathbf{C}(t).$$

For $\mathbf{W}$ near $\mathbf{Y}$ we can approximate $\mathbf{y}(t; \mathbf{W})$ by

$$\mathbf{y}(t; \mathbf{Y}) + \mathbf{C}(t; \mathbf{Y})(\mathbf{W} - \mathbf{Y}).$$

We therefore define a transformation to connection coordinates $\mathbf{z}(t)$ by

$$\mathbf{y}(t; \mathbf{W}) = \mathbf{y}(t; \mathbf{Y}) + \mathbf{C}(t; \mathbf{Y})\mathbf{z}(t; \mathbf{Y}, \mathbf{W}).$$

Because $\mathbf{C}(0) = \mathbf{I}$ and $\mathbf{C}' = \mathbf{J}\mathbf{C}$, the transformed equation for $\mathbf{z}$ has the form

$$\mathbf{z}' = \mathbf{C}(t)^{-1}\{\mathbf{f}(\mathbf{y}(t; \mathbf{Y}) + \mathbf{C}(t)\mathbf{z}(t)) - \mathbf{f}(\mathbf{y}(t; \mathbf{Y}))\} - \mathbf{C}(t)^{-1}\mathbf{J}(t)\mathbf{C}(t)\mathbf{z}.$$

EXERCISE

Derive the above differential equation for $\mathbf{z}$.

Additional reference: 80.

10.5 GENERAL COMPUTATIONAL PROBLEMS

As presented, all of the methods given so far in this chapter assume that we have the solution $\mathbf{y}(t;\, \mathbf{Y})$ in hand; of course, this is seldom true, although we may expect to have an *approximate* solution $\mathbf{y}^*(t;\, \mathbf{Y})$. As mentioned in Section 10.3, we would then, in practice, define the transformation to new coordinates by

$$\mathbf{y}(t;\, \mathbf{W}) = \mathbf{y}^*(t;\, \mathbf{Y}) + \mathbf{\Gamma}(t;\, \mathbf{y}^*)\mathbf{z}(t;\, \mathbf{Y},\, \mathbf{W}),$$

where $\mathbf{\Gamma}(t;\, \mathbf{y}^*)$ is one of the matrices $\mathbf{E}(t)$, $\mathbf{V}(t)$, $\mathbf{C}(t)$, or an approximation thereto. In the differential equations derived for $\mathbf{z}(t)$ one need then replace $\mathbf{f}(t,\, \mathbf{y})$ by $\mathbf{y}^{*\prime}$. We should point out that many transformations other than those we have discussed have been suggested; in essence, these involve a different choice of $\mathbf{\Gamma}(t;\, \mathbf{y})$. In particular, transformations which yield ellipsoidal error bounds or probabilistic error estimates have been studied; these are essentially of the type we have discussed, although they differ in detail and effect.

We must remark that in computing $\mathbf{z}(t;\, \mathbf{Y},\, \mathbf{W})$ from its differential equation, one must be careful about the loss of accuracy caused by subtraction of nearly equal numbers on the right-hand side of that differential equation. If possible some of these operations should be performed analytically or, better still, automatically by computer-symbol manipulation routines. Only in this way can the full advantage of reducing the rotation of the system be attained.

10.6 A COMPUTATIONAL PROCEDURE USING CONNECTION COORDINATES

As we remarked in the preceding section, in practice we would compute $\mathbf{y}(t;\, \mathbf{Y})$ and $\mathbf{\Gamma}(t;\, \mathbf{y})$ approximately. For clarity we indicate a procedure that uses connection coordinates; to keep matters simple, let us consider the first-order Taylor-series method known as Euler's method. Let us denote by $\mathbf{Y}_i$, $\mathbf{W}_i$, $\mathbf{C}_i$, and $\mathbf{Z}_i$ numerical approximations to $\mathbf{y}(t_i;\, \mathbf{Y})$, $\mathbf{y}(t_i;\, \mathbf{W})$, $\mathbf{C}(t_i;\, \mathbf{Y})$ and $\mathbf{z}(t_i;\, \mathbf{Y},\, \mathbf{W})$, respectively, where, for convenience, we set $t_i = ih$.

If we have computed the numbers $\mathbf{Y}_i$, then we can evaluate the Jacobian matrix $\mathbf{J}(\mathbf{Y}_i)$ so that Euler's method for computing $\mathbf{C}_i$ becomes

$$\mathbf{C}_{i+1} = \mathbf{C}_i + h\mathbf{J}(\mathbf{Y}_i)\mathbf{C}_i,$$

and Euler's method for computing $\mathbf{z}_i$ becomes

$$\mathbf{Z}_{i+1} = \mathbf{Z}_i + h\mathbf{C}_i^{-1}\{\mathbf{f}(\mathbf{Y}_i + \mathbf{C}_i\mathbf{Z}_i) - \mathbf{Y}_i' - \mathbf{J}(\mathbf{Y}_i)\mathbf{C}_i\mathbf{Z}_i\}.$$

Combining these two formulas and using the fact that

$$\mathbf{W}_i = \mathbf{Y}_i + \mathbf{C}_i\mathbf{Z}_i,$$

we conclude that

$$\mathbf{W}_{i+1} = \mathbf{W}_i + \mathbf{Y}_{i+1} - \mathbf{Y}_i + h(\mathbf{f}(\mathbf{W}_i) - \mathbf{Y}_i') + h^2\mathbf{J}(\mathbf{Y}_i)[\mathbf{f}(\mathbf{W}_i) - \mathbf{Y}_i' - \mathbf{J}(\mathbf{Y}_i)(\mathbf{W}_i - \mathbf{Y}_i)].$$

In this form we have a numerical method for computing $\mathbf{W}_i \sim \mathbf{y}(t_i; \mathbf{W})$ without directly using $\mathbf{C}_i \sim \mathbf{C}(t_i; \mathbf{Y})$.

EXERCISE

Apply these formulas in a numerical computation for the example discussed in Section 10.7 using $h = 0.1$, $i = 1, 2, \ldots, 60$. Compare results with the exact solution and with results obtained using Euler's method (also with $h = 0.1$), directly on equations $y_1' = y_2$, $y_2' = -4y_1$, $y_1(0) = 0$, $y_2(0) = 2$.

10.7 GEOMETRIC VIEW OF THE TRANSFORMATIONS

To give a better intuitive idea of the effect of these transformations, we consider their application to the trivial linear system

$$y_1' = y_2$$

$$y_2' = -4y_1.$$

The trajectory through

$$\mathbf{Y} = \begin{pmatrix} 0 \\ 2 \end{pmatrix}$$

is

$$y_1(t) = \sin 2t$$

$$y_2(t) = 2 \cos 2t;$$

we examine the trajectory

$$w_1(t) = 1.1 \sin 2t + 0.1 \cos 2t$$

$$w_2(t) = 2.2 \cos 2t - 0.2 \sin 2t$$

through the nearby point

$$\mathbf{W} = \begin{pmatrix} 0.1 \\ 2.2 \end{pmatrix}.$$

The trajectories are shown in Figure 10–2.

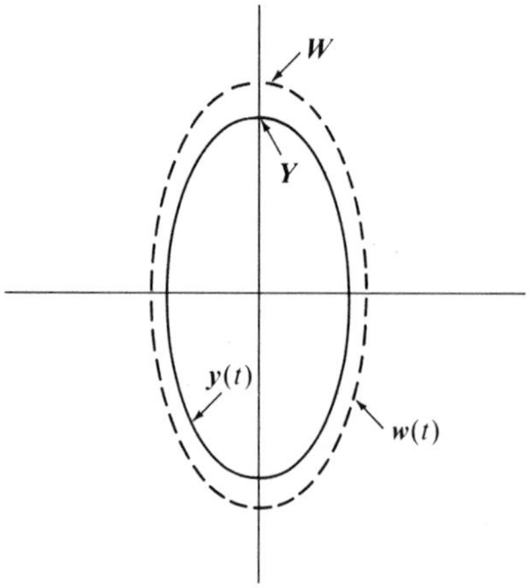

FIGURE 10–2

EXAMPLE Intrinsic coordinates. In intrinsic coordinates, the point $\mathbf{y}(t; \mathbf{W}) = \mathbf{w}(t)$ is described via the components of $\mathbf{w}(t) - \mathbf{y}(t; \mathbf{Y}) = \mathbf{w}(t) - \mathbf{y}(t)$ in the orthogonal coordinate system with origin at $\mathbf{y}(t)$ and axes tangent and normal to the trajectory. The $\mathbf{z}$-coordinate axes at $t = \pi/8$ are shown in Figure 10–3.

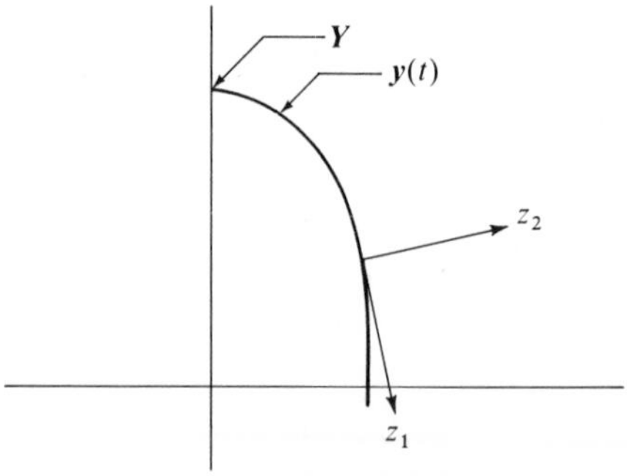

FIGURE 10–3

If we perform the transformation and solve for $\mathbf{z}(t)$, we find that

$$z_1(t) = \frac{1}{\sqrt{1 + 3\sin^2 2t}}\left[-0.3\sin 2t\cos 2t + 0.1(1 + 3\sin^2 2t)\right]$$

$$z_2(t) = \frac{0.2}{\sqrt{1 + 3\sin^2 2t}}.$$

In $\mathbf{z}$ coordinates the trajectory has been greatly reduced in size; it is sketched in Figure 10–4 with a magnified scale. ●

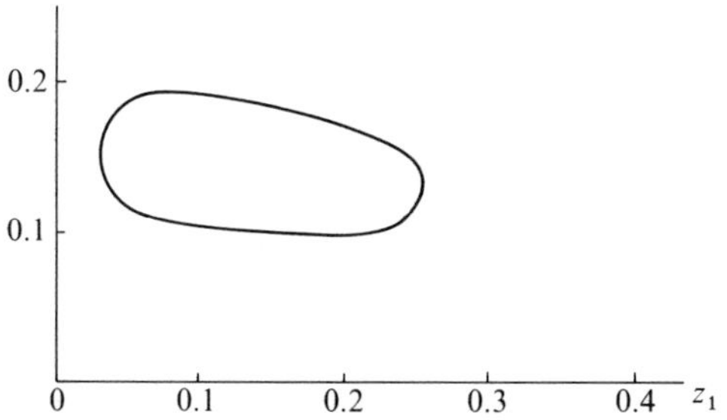

FIGURE 10–4

EXAMPLE Linearized and connection coordinates. Since the equation is linear, both of these methods provide the same transformation to $\mathbf{z}$ coordinates given by

$$\mathbf{w}(t) = \mathbf{y}(t) + \begin{pmatrix} \cos 2t & \frac{1}{2}\sin 2t \\ -2\sin 2t & \cos 2t \end{pmatrix}\mathbf{z}(t).$$

The unit-coordinate vectors at $\pi/8$ are shown in Figure 10–5.

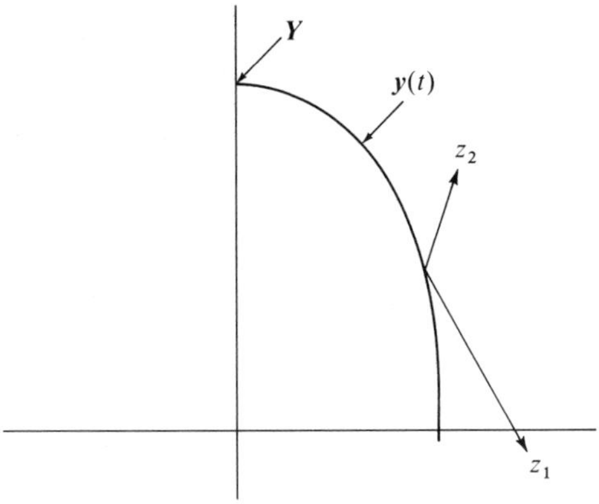

FIGURE 10–5

In these coordinates the flow has stopped completely, that is

$$z_1(t) = 0.1,$$

$$z_2(t) = 0.2. \quad \bullet$$

10.8 USE AS A BASIS FOR DESIGNING PREDICTOR-CORRECTOR METHODS

We have discussed the use of these transformations to solve a system with varying initial conditions; they can also be used to correct a previously predicted solution through a fixed initial point $\mathbf{Y}$. That is, given an approximate solution $\mathbf{y}^*(t)$, we can base a transformation on it, compute $\mathbf{z}(t)$ in some fashion in the transformed coordinates, and transform back, yielding a corrected solution $\mathbf{y}(t)$.

EXAMPLE We consider again the nonlinear system

$$y_1' = y_2$$

$$y_2' = -\sin y_1,$$

$$y_1(0) = 0, \qquad y_2(0) = 20.$$

Let us take as an approximate solution $\mathbf{y}^*(t)$ the first terms of the Taylor series at $t = 0$,

$$\mathbf{y}^*(t) = \mathbf{y}(0) + t\mathbf{y}'(0) + \frac{t^2}{2}\mathbf{y}''(0) + \frac{t^3}{6}\mathbf{y}'''(0)$$

$$= \begin{pmatrix} 20t - \dfrac{10t^3}{3} \\ 20 - 10t^2 \end{pmatrix}.$$

Now let $\mathbf{Y}^*(t)$ be the matrix with columns given by $\mathbf{y}^{*\prime}(t)$, $\mathbf{y}^{*\prime\prime}(t)$ and define new linearized $\mathbf{z}$ coordinates by

$$\mathbf{y}(t) = \mathbf{y}^*(t) + \mathbf{Y}^*(t)\mathbf{Y}^*(0)^{-1}\mathbf{z}(t).$$

The functions $z_1(t)$ and $z_2(t)$ satisfy

$$z_1' = \frac{t}{1 + t^2/2}\left\{\sin\left[\frac{20t - t^3}{6} + \frac{1 - t^2}{2}\, z_1 + tz_2\right] - 20t - z_1\right\}$$

$$z_2' = \frac{1 - t^2/2}{1 + t^2/2}\left\{20t + z_1 - \sin\left[\frac{20t - t^3}{6} + \frac{1 - t^2}{2}\, z_1 + tz_2\right]\right\}.$$

EXERCISE

Verify.

We can now solve the equations for $\mathbf{z}$ rather than for $\mathbf{y}$. Since

$$\mathbf{z}(0) = \mathbf{z}'(0) = \mathbf{z}''(0) = \mathbf{z}'''(0) = 0,$$

the $\mathbf{z}$ trajectories will change slowly for small t, but not necessarily for a large range of t values; therefore in implementing this transformation, it would be best to use it only over a small t interval before computing a new Taylor expansion and thereby a new transformation. The t intervals over which each expansion and transformation is used can be determined automatically by machine [80, 81].

To demonstrate that the transformation above has reduced the flow numerically, we exhibit in Figure 10–6 the values of $y_1(t)$, $y_2(t)$, $z_1(t)$, and $z_2(t)$ over a small t interval. ●

t	y_1	y_2	z_1	z_2
0.0	0	20.0	0	0
0.02	0.39997 . . .	19.99605 . . .	−0.0000008 . . .	0.004 . . .
0.04	0.79979 . . .	19.98484 . . .	−0.000027 . . .	0.017 . . .
0.06	1.19933 . . .	19.96812 . . .	−0.0002 . . .	0.04 . . .
0.08	1.59850 . . .	19.94854 . . .	−0.0008 . . .	0.08 . . .
0.10	1.99727 . . .	19.92919 . . .	−0.002 . . .	0.13 . . .

FIGURE 10–6

BIBLIOGRAPHY

1. Abramowitz, M. and I. Stegun, *Handbook of mathematical functions*, New York, Dover, 1965.
2. Akhiezer, N., *The calculus of variations*, Waltham, Blaisdell, 1962.
3. Anselone, P., *Nonlinear integral equations*, Madison, Wisconsin Press, 1964.
4. Babuska, I., M. Prager, and E. Vitasek, *Numerical processes in differential equations*, New York, Interscience, 1966.
5. Babuska, I. and S. Sobolev, "Optimizatsiya chislennych metodov," *Aplik. Mat.*, vol. 10 (1965), 96–129 (Russian).
6. Bellman, R., *Perturbation techniques in mathematics, physics, and engineering*, New York, Holt, 1966.
7. Bellman, R., *Stability theory of differential equations*, New York, McGraw-Hill, 1953.
8. Bellman, R. and R. Kalaba, *Quasilinearization and nonlinear boundary value problems*, New York, Elsevier, 1965.
9. Birkhoff, G., C. de Boor, B. Swartz, and B. Wendroff, "Rayleigh-Ritz approximation by piecewise cubic polynomials," *SIAM J. Num. Anal.*, vol. 3 (1966), 188–203.
10. Birkhoff, G. and G-C Rota, *Ordinary differential equations*, Boston, Ginn, 1962.
11. Braun, J. and R. Moore, "Solution of differential equations," *Math. Res. Cntr. Report* #901 (1968), Madison.
12. Brouke, M., Universite Louvain, private communication.
13. Brunner, H., "Stabilization of optimal difference operators," *Z. Angew. Math. Phys.*, vol. 18 (1967), 438–444.
14. Bulirsch, R. and J. Stoer, "Numerical treatment of ordinary differential equations by extrapolation methods," *Num. Math.*, vol. 8 (1966), 1–13.
15. Butcher, J., "A modified multistep method for the numerical integration or ordinary differential equations," *JACM*, vol. 12 (1965), 124–135.
16. Butcher, J., "Implicit Runge-Kutta processes," *Math. Comp.*, vol. 18 (1964), 50–64.
17. Butcher, J., "On the integration processes of A. Huta," *J. Australian Math. Soc.*, vol. 3 (1963), 202–206.

18. Cesari, L., *Asymptotic behavior and stability problems in ordinary differential equations*, Berlin, Springer, 1959.
19. Chaplygin, S., "Approximate integration of a system of two differential equations," Private translation from his *Collected Works*, Volume II, Gostekhidzat, 1948 (Russian).
20. Cherry, T., "Integrals of systems of ordinary differential equations," *Proc. Comb. Phil. Soc.*, vol. 22 (1924), 273–281.
21. Ciarlet, P., M. Schultz, and R. Varga, "Numerical methods of high order accuracy for nonlinear boundary value problems: I. One dimensional problem," *Num. Math.*, vol. 9 (1967), 394–430.
22. Ciarlet, P., M. Schultz, and R. Varga, "Numerical methods of high order accuracy for nonlinear boundary value problems: II. Nonlinear boundary conditions," *Num. Math.*, vol. 11 (1968), 331–345.
23. Ciarlet, P., M. Schultz, and R. Varga, "Numerical methods of high order accuracy for nonlinear boundary value problems: V. Monotone operator theory," *Num. Math.*, vol. 13 (1969), 51–77.
24. Coddington, E., Levinson, N., *Theory of ordinary differential equations*, New York, McGraw-Hill, 1955.
25. Collatz, L., *Functional analysis and numerical mathematics*, New York, Academic, 1966.
26. Collatz, L., "Monotonicity and related methods in nonlinear differential equation problems," *In* D. Greenspan editor, *Numerical solution of nonlinear differential equations*, New York, Wiley, 1966, pp. 65–87.
27. Collatz, L., *The numerical treatment of differential equations*, Berlin, Springer, 1960.
28. Conley, C., University of Wisconsin, Madison, private communication.
29. Courant, R. and D. Hilbert, *Methods of mathematical physics*, vol. I, New York, Interscience, 1953.
30. Dahlquist, G., "A numerical method for some ordinary differential equations with large Lipschitz constants," *Proc. IFIP Cong. 1968, Supp.*, Book. I, pp. 32–36.
31. Dahlquist, G., "A special stability problem for linear multistep methods," *BIT*, vol. 3 (1963), 27–43.
32. Dahlquist, G., "Convergence and stability in the numerical integration of ordinary differential equations," *Math. Scand.*, vol. 4 (1956), 33–53.
33. Dahlquist, G., "On rigorous error bounds in the numerical solution of ordinary differential equations," in D. Greenspan editor, *Numerical solutions of nonlinear differential equations*, New York, Wiley, 1966, pp. 89–95.
34. Dahlquist, G., "Stability questions for some numerical methods for ordinary differential equations," *Proc. Symp. Appl. Math.*, vol. 15 (1963), 147–158.
35. Daniel, J., "Nonlinear equations arising in deferred correction of initial value problems," *Math. Res. Cntr. Report* #818 (1967), Madison. Also in *Acta Cient. Venezolana*, vol. 19 (1968), 123–128.
36. Daniel, J., "On the approximate minimization of functionals," *Computer Sci. Report* #42 (1968), Univ. of Wisconsin, Madison. Also in *Math. Comp.*, vol. 23 (1969).
37. Daniel, J., V. Pereyra, and L. Schumaker, "Iterated deferred corrections for initial value problems," *Math. Res. Cntr. Report* #808 (1967), Madison. Also in *Acta Cient. Venezolana*, vol. 19 (1968), 128–135.

38. Davis, H., *Introduction to nonlinear differential and integral equations*, New York, Dover, 1962.
39. Durham, H., Jr., et al., "Study of methods for numerical solution of ordinary differential equations," *Final report, NAS8–11129, Project A-740*, Atlanta, Georgia Institute of Technology, 1964.
40. Engeli, M., Th. Ginsburg, H. Rutishauser, and E. Stiefel, *Refined iterative methods for computation of the solution and the eigenvalues of self adjoint boundary value problems, Mitt. Inst. Angew. Math.* #8, Basel, 1959.
41. Fehlberg, E., "Neue genauere Runge-Kutta Formeln fur Differential-gleichungen *n*-ter Ordnung," *Z. Angew. Math. Mech.*, vol. 40 (1960), 449–455.
42. Fehlberg, E., "New high order Runge-Kutta formulas with step size control for systems of first and second order differential equations," *Z. Angew. Math. Mech.*, vol. 44 (1964), 17–29.
43. Forsyth, A., *Theory of differential equations*, London, Cambridge University Press, 1900.
44. Fox, L., *Numerical solution of ordinary and partial differential equations*, Reading, Pa., Addison-Wesley, 1962.
45. Fox, L., *The numerical solution of two point boundary problems in ordinary differential equations*, London, Oxford University Press, 1957.
46. Gallaher, L. and I. Perlin, "A comparison of several methods of numerical integration of nonlinear differential equations," Address at SIAM meeting, University of Iowa, March 11–14, 1966.
47. Gear, C. W., "Hybrid methods for initial value problems in ordinary differential equations," *SIAM J. Num. Anal.*, vol. 2 (1965), 69–86.
48. Gear, C. W., "The automatic integration of stiff ordinary differential equations," *Proc. IFIP Cong. 1968, Math.*, Book. A, 81–85.
49. Gear, C. W., "The numerical integration of ordinary differential equations," *Math. Comp.*, vol. 21 (1967), 146–156.
50. Gragg, W., "On extrapolation algorithms for ordinary initial value problems," *SIAM J. Num. Anal.*, vol. 2 (1965), 384—403.
51. Gragg, W. and H. Stetter, "Generalized multistep predictor-corrector methods," *JACM*, vol. 11 (1964), 188–209.
52. Greenspan, D., "Approximate solution of initial value problems for ordinary differential equations by boundary value techniques," *Math. Res. Cntr. Report* #752 (1967), Madison. Also in *CACM*, vol. 11 (1968), 760–763.
53. Groebner, W. and H. Knapp, *Contributions to the method of Lie series*, Bibliographisches Institut Mannheim, 1967.
54. Hansen, E., "Interval bounds for the solution of two point boundary value problems," Address to Office of Naval Research Summer Conference on Approximation, Cornell, 1968. Also in *Topics in Interval Analysis*, Oxford University Press, 1970.
55. Henrici, P., *Discrete variable methods in ordinary differential equations*, New York, Wiley, 1962.
56. Henrici, P., *Error propagation for difference methods*, New York, Wiley, 1963.
57. Hull, T., "A search for optimum methods for the numerical integration of ordinary differential equations," *SIAM Rev.*, vol. 9 (1967), 647–654.
58. Hull, T., "The numerical integration of ordinary differential equations," *Computer Sci. Report* #1 (1968), Univ. of Toronto. Also in *Proc. IFIP Cong. 1968.*

59. Hull, T. and R. Johnston, "Optimum Runge-Kutta methods," *Math. Comp.*, vol. 18 (1964), 306–310.
60. Hull, T. and A. Newberry, "Integration procedures which minimize propagated errors," *J. SIAM*, vol. 9 (1961), 31–47.
61. Ince, E., *Ordinary differential equations*, New York, Dover, 1944.
62. Jackson, L., "A comparison of ellipsoidal and interval arithmetic error bounds," Lecture presented at SIAM Fall meeting, Philadelphia, 1968.
63. Kahan, W., "A computable error bound for systems of ordinary differential equations," *SIAM Rev.*, vol. 8 (1966), 568–569.
64. Keller, H., *Numerical methods for two point boundary value problems*, Waltham, Blaisdell, 1968.
65. Knapp, H., "Uber eine Verallgemeinerung des Verfahrens der sukzessiven Approximation zur Losung von Differentialgleichungsystemen," *Monatshefte fur Math.*, vol. 68 (1964), 33–45.
66. Knapp, H. and G. Wanner, "LIESE, a program for ordinary differential equations using Lie series," *Math. Res. Cntr. Report* #881 (1968), Madison.
67. Knapp, H. and G. Wanner, "Numerical solution of ordinary differential equations by Groebner's Lie series method," *Math. Res. Cntr. Report* #880 (1968), Madison.
68. Krabs, W., "Einige Methoden zur Losung des diskreten linearen Tschebyscheff-Problems," Dissertation, Univ. of Hamburg (1963).
69. Krogh, F., "A variable step variable order multistep method for the numerical solution of ordinary differential equations," *Proc. IFIP Cong. 1968, Math.*, Book. A, 91–95.
70. Lambert, J. and B. Shaw, "A generalization of multistep methods for ordinary differential equations," *Num. Math.*, vol. 8 (1966), 250–263.
71. Lambert, J. and B. Shaw, "On the numerical solution of $y' = f(x,y)$ by a class of formulae based on rational approximation," *Math. Comp.*, vol. 19 (1965), 456–462.
72. Lawson, J., "Generalized Runge-Kutta processes for stable systems with large Lipschitz constants," *SIAM J. Num. Anal.*, vol. 4 (1967), 372–380.
73. Lees, M., "Discrete methods for nonlinear two point boundary value problems," *In* J. Bramble, editor, *Numerical solution of partial differential equations*, New York, Academic, 1966, pp. 59–72.
74. Lefschetz, S., *Differential equations, geometric theory*, New York, Interscience, 1957.
75. Loscalzo, F. and I. Schoenberg, "On the use of spline functions for the approximation of solutions of ordinary differential equations," *Math. Res. Cntr. Report* #723 (1967), Madison.
76. Loscalzo, F. and T. Talbot, "Spline function approximations for solutions of ordinary differential equations," *SIAM J. Num. Anal.*, vol. 4 (1967), 433–445.
77. Lotkin, M., "On the accuracy of Runge-Kutta's method," *MTAC*, vol. 5 (1951), 128–132.
78. McCuskey, S., *Introduction to celestial mechanics*, Reading, Addison-Wesley, 1963.
79. Miranker, W. and W. Liniger, "Parallel methods for the numerical integration or ordinary differential equations," *Math. Comp.*, vol. 21 (1967), 303–320.
80. Moore, R., "Automatic local coordinate transformations to reduce the growth of error bounds in interval computation of solutions of ordinary

differential equations," *In* L. Rall, editor, *Error in digital computation*, vol. II, New York, Wiley, 1965, pp. 103–140.
81. Moore, R., *Interval analysis*, Englewood Cliffs, Prentice-Hall, 1966.
82. Moore, R., "Practical aspects of interval computation," *Aplik. Mat.*, vol. 13 (1968), 52–92.
83. Nordsieck, A., "On numerical integration of ordinary differential equations," *Math. Comp.*, vol. 16 (1962), 22–49.
84. Ortega, J. and M. Rockoff, "Nonlinear difference equations and Gauss-Seidel type iterative methods," *SIAM J. Num. Anal.*, vol. 3 (1966), 497–513.
85. Osborne, M., "A new method for the integration of stiff systems of ordinary differential equations," *Proc. IFIP Cong. 1968, Math.*, Book. A, pp. 86–90.
86. Ostrowski, A., *Solution of equations and systems of equations*, New York, Academic, 1960.
87. Pereyra, V., "Iterated deferred corrections for nonlinear boundary value problems," *Num. Math.*, vol. 11 (1968), 111–126.
88. Pereyra, V., "Iterated deferred corrections for nonlinear operator equations," *Num. Math.*, vol. 10 (1967), 316–323.
89. Rabinowitz, P., "Applications of linear programming to numerical analysis," *SIAM Rev.*, vol. 10 (1968), 121–160.
90. Rosen, J. and R. Meyer, "Solution of nonlinear two point boundary value problems by linear programming," *Computer Sci. Report* #1 (1967), Univ. of Wisconsin, Madison.
91. Rosser, J., "A Runge-Kutta for all seasons," *SIAM Rev.*, vol. 9 (1967), 417–452.
92. Scraton, R., "Estimation of the truncation error in Runge-Kutta and allied processes," *Computer J.*, vol. 7 (1964), 246–248.
93. Shampine, L., "Monotone iterations and two sided convergence," *SIAM J. Num. Anal.*, vol. 3 (1966), 607–615.
94. Shanks, E., "Formulas for obtaining solutions of differential equations by evaluations of functions," Address at AMS Summer Meeting, Boulder, 1963.
95. Shaw, B., "Modified multistep methods based on a nonpolynomial interpolant," *JACM*, vol. 14 (1967), 143–154.
96. Stewart, N., "The comparison of numerical methods for ordinary differential equations," *Computer Sci. Report* #3 (1968), Univ. of Toronto.
97. Stromgren, E., *Connaissance actuelle des orbites dans le problem des trois corps*, Memoires et Varietes, T.IX, Copenhagen Observatory, 1935.
98. Sylvester, R. and F. Meyer, "Two point boundary problems by quasilinearization," *J. Appl. Math.*, vol. 13 (1965), 586–602.
99. Talbot, T., "Guaranteed error bounds for computed solutions of nonlinear two point boundary value problems," *Math. Res. Cntr. Report* #875 (1968), Madison.
100. Urabe, M., "Galerkin's procedure for nonlinear periodic systems and its extension to multipoint boundary value problems for general nonlinear systems," *In* D. Greenspan, editor, *Numerical solutions of nonlinear differential equations*, New York, Wiley, 1966, pp. 297–327.
101. Urabe, M., and A. Reiter, "Numerical computation of nonlinear forced oscillations by Galerkin's method," *Math. Res. Cntr. Report* #510 (1964), Madison.
102. Urabe, M., H. Yanagiwara, and Y. Shinohara, "Periodic solutions of vander

Pol's equation with damping coefficient $\lambda = 2 \sim 10$," *J. Sc. Hirosh.*, vol. 23 (1960), 325–366.

103. Van Wyk, R., "Variable mesh multistep methods for ordinary differential equations," *Rocketdyne Res. Report* #67–12 (1967).
104. Varga, R., "Hermite interpolation type methods for two point boundary value problems," *In* J. Bramble, editor, *Numerical solution of partial differential equations*, New York, Academic, 1965, pp. 365–373.
105. Varga, R., *Matrix iterative analysis*, Englewood Cliffs, Prentice-Hall, 1962.
106. Widlund, O., "A note on unconditionally stable linear multistep methods," *BIT*, vol. 7 (1967), 65–70.
107. Willmore, T., *Introduction to differential geometry*, London, Oxford University Press, 1959.
108. Wintner, A., *The analytical foundations of celestial mechanics*, Princeton, Princeton University Press, 1941.
109. Zonneveld, J., "Automatic numerical integration," *Mathematical Centre Tracts* #8, 1964, Amsterdam.

INDEX